The Hidden Evil

The Financial Elite's Covert War Against the Civilian Population

Mark M. Rich

Published by Lulu Enterprises, Inc.

Published by
Lulu Enterprises, Inc.
860 Aviation Parkway, Suite 300
Morrisville, NC 27560
www.lulu.com

Printed in the United States of America
First Paperback Edition 2008
ISBN 978-1-4357-5010-4
For book orders:
www.lulu.com
www.amazon.com
www.barnesandnoble.com

Contents

Volume I Part I

Volume I Part II

Volume II Part I

Volume II Part II

Volume I Part I

Chapter 1

Overview

Throughout this book I will be presenting evidence on what I've discovered to be highly organized, covert, psychological warfare campaigns being carried out on the civilian population in all NATO countries. Upon researching this phenomenon, I found that it led to elite organizations that have instigated wars, manipulated major historical events, and have wreaked havoc on a population that is largely unaware of their existence. After reading this book you may draw your own conclusions.

I will be explaining something that resembles insanity, but isn't insanity, yet is caused by those who are insane. This will become evident later. The information in this book will be uncomfortable to most people. Some of it will be scary. I do not present the information for the purpose of scaring you, but it is a peripheral result of explaining something this repulsive in detail. It appears that the more areas of a horrific phenomenon you attempt to illuminate, the more you discover how horrifying it really is.

There are multiple websites now devoted to this topic which are listed in the Resources section of this book. Some independent talk shows such as *The Grassy Knoll*, *The Power Hour*, *The Investigative Journal*, and a few others, have covered this topic. But it is not yet recognized by the mainstream media, the legal community, or mental health professionals, for reasons which I will explain.

Some of the people I'll be referencing have been labeled conspiracy nuts by the mainstream media in order to be discredited.(*) Interestingly, I've found that some of those engaged in the paranormal are partially responsible for the creation of this program. So, as I'll demonstrate, the financial elite recruit the services of these individuals when required, but attack them when they expose their operations. These are standard discrediting tactics. This will be explored in the *Tactics* and *Techniques to Discredit* chapters.

Each person or organization used as a source may not necessarily agree with my overall estimation of this, or with other sources listed. They are used to illustrate specific points. If you're on the receiving end of this policy, or even if you're not targeted but have researched it, you may not completely agree with my findings. I'm just presenting the information that I've compiled. I don't propose to have the absolute truth regarding this phenomenon, but my aim is to obtain it.

There are people who will have a vested interest in discrediting me or the information in this book. I've taken much caution to minimize the amount of informational errors that it contains. But most books contain some errors.(**) I have taken little liberty when making certain points that I believe are pertinent by doing my best to document this phenomenon as much as possible.

But, as much as I've done my best to do this, the media, and "experts," may find something wrong with it and begin a relentless campaign to discredit me. Some words or phrases may be twisted around and quoted out of context. Small irrelevant inconsistencies may be blown out of proportion. There may also be a pervasive effort by multiple respectable "victims" to discredit me, or this information. As I'll demonstrate, these are predictable tactics. But keep this in mind—after absorbing the information, ask yourself; *does it ring true to you*?

I will be referencing the work of doctors, former federal law enforcement, politicians, and researchers. Some of these individuals have seen how this planet really works due to their experience on the front lines. This experience has given them a glimpse of matters that will eventually affect everyone. They have issued warnings to the public in their books and publications. But apparently too many are asleep.

During this presentation you'll see evidence from both the political *Left* and the *Right*. Because there are often barricades which prevent people who have adopted a political side from receiving information, I ask that you briefly abandon any identification with a political party. This is necessary for two reasons. First, mainstream conservative and liberal news sources both occasionally publish useful information. Second, as I'll demonstrate, both political parties are controlled by the same source. The *left/right* political spectrum is a control mechanism.

Acknowledging this trap in his book, *America's Secret Establishment*, Professor Antony C. Sutton stated, "Above all the reader must—at least temporarily while reading this work—put to one side the descriptive clichés of the left and right, liberal and conservative, communist and fascist, even republican and democrat. These terms may be important for self-recognition, they do provide a certain reassurance, but they are confusing in our context unless seen as essential elements in a game plan."

The first part of this volume (Volume I), explains how the Hidden Evil can be allowed to exist in our society. This volume will cover the controlling financial elite, who they are, what they've done, and what their plans are. It will provide evidence to support my premise that, unknown to the masses, elite groups control their governments.

The second part of Volume I is an overview of historical events that have been kept from mainstream publications. It provides an outline which explains how the Hidden Evil is part of a recurring historical pattern. It also explains how these elite groups control the media, schools, and have rewritten history. Furthermore, it documents their financial support for Russian and German dictatorships, as well as a string of lies they've used to cover their tracks.

Volume II will focus on the structure, tactics, and purpose for the Hidden Evil, as well as how people can be persuaded to serve it. Also in Volume II, we'll explore the high probability that it was created by these elite organizations. I'll also cover the subject of Satanism which has surfaced many times during my research, and manifests itself on multiple levels within this program. But, it is not necessary to believe in Satan, Christ, or any religion in order to recognize that the Hidden Evil exists.

This book is not a complete coverage of the Global Union (New World Order), or of the global financial elite. Neither is it an in-depth coverage of history, or Non-lethal Weapons (NLW), also called Directed Energy Weapons (DEW). But it's necessary that these subjects are touched upon to show a pattern which suggests that the Hidden Evil is a foundational element for the maintenance of a dictatorship, and was created by people who have funded dictatorships in the past.

Why would I want to focus on such a dark area of humanity? There is the argument by some mystics or *New Agers* that much of reality is an illusion. They insist that perception is reality, and if you choose to change your interpretation of an event, then it will have whatever meaning you assign to it. I think that is the case. Many things can be changed by choosing a different belief system. But I also think there are principles and circumstances which exist no matter what. They exist regardless of our beliefs or knowledge of their existence. The Hidden Evil is one of them.

Endnotes

* Dr. Rauni Leena Kilde, who has written about this phenomenon, has been labeled a conspiracy nut (wikipedia.com), for believing that there may be intelligent life in the universe besides human beings.

** This book will contain grammatical and spelling errors. Some of them are my responsibility. In addition, the manuscript for this book has undergone frequent sabotage. This occurred from the time I began writing it, up until (and including) the time it was uploaded to the publisher's server.

Chapter 2

Introduction to the Financial Elite

Before covering the specific groups which are likely responsible for creating the Hidden Evil, it's necessary to explain that the United States and other NATO countries are not run by their respective leaders. The power structure does not operate in the manner taught in public schools. The evidence I'll be presenting suggests that these nations are run by an elite cabal of wealthy individuals who prefer to operate from behind the scenes. This group has been called the *Establishment*, the *Secret Brotherhood*, the *Alliance*, the *Invisible Government*, the *Shadow Government*, a *legal mafia*, the *New World Order*, the *Illuminati*, the *Insiders*, the *Military-industrial Complex*, etc.(*)

In an article entitled, *Elite Clique Holds Power in the U.S.*, which appeared in the *Indianapolis News* on December 23, 1961, Edith Kermit Roosevelt wrote, "The word 'Establishment' is a general term for the power elite in international finance, business, the professions and government, largely from the northeast, who wield most of the power regardless of who is in the White House. Most people are unaware of the existence of this 'legitimate Mafia.'" She described, "The power of the Establishment makes itself felt from the professor who seeks a foundation grant, to the candidate for a cabinet post or State Department job. It affects the nation's policies in almost every area."

Most people can sense that something is wrong, but can't quite put their finger on what it is. Despite changes in administration every four years, the planet continues to fall apart. Professor Sutton stated in his book, *Wall Street and the Rise of Hitler*, "Wars are started (and stopped) with no shred of coherent explanation. Political words have never matched political deeds." The reason, he asserted, is "because the center of political power has been elsewhere than with elected and presumably responsive representatives in Washington, and this power elite has its own objectives, which are inconsistent with those of the public at large."

On March 26, 1922, the Mayor of New York City, John F. Hylan said, "The real menace of our republic is the invisible government which, like a giant octopus, sprawls its slimly length over our city, state and nation. At the head is a small group of banking houses generally referred to as 'international bankers.' This little coterie of powerful international bankers virtually run our government for their own selfish ends."

Congressman Lawrence P. McDonald wrote in November of 1975, "Money alone is not enough to quench the thirst and lusts of the super-rich. Instead, many of them use their vast wealth, and the influence such riches give them, to achieve even more power. Power of a magnitude never dreamed of by the tyrants and despots of earlier ages. Power on a worldwide scale. Power over people, not just products." In his book *Dark Majesty*, Professor Texe Marrs added, "The men at the top of the empire are billionaires many times over. Yet they still want more money."

Author James Perloff describes the typical path of one of these individuals, in his book, *The Shadows of Power*. He writes that it "begins in private schools, the most famous being Groton. From these they have typically proceeded to Harvard, Yale, Princeton, or Columbia, there entering exclusive fraternities, such as Yale's secretive Skull and Bones. ... From academia they have customarily progressed to Wall Street, perhaps joining an international investment bank, such as Chase Manhattan, or a prominent law firm or brokerage house."

He continues, "Some of the politically inclined have signed on with the establishment think tanks like the Brookings Institution and the Rand Corporation," and "a few have found themselves on the boards of vast foundations." From there, states Perloff, they ascend into "high positions in the federal government," which requires membership in a "New York-based group called the Council on Foreign Relations."

Professor Marrs declared, "Five major groups—Banking and Money, Political, Intelligence, Religious, and Educational—are under the immediate direction and control of this small band of men." Everybody has heard of the "they" who seem to be in control of things. "Whatever we call this self-perpetuating elitist group," explained Professor Sutton, "it is apparently fundamentally significant in the determination of world affairs, at a level far behind and above that of the elected politicians."

Before we explore who they are, I should point out that throughout this book you'll notice a steady pattern of the following:

- They disguise themselves and their intentions by creating groups which profess only the most innocent objectives. In this manner, they masquerade as humanitarians.

- They frequently resort to lying and deceiving to accomplish their objectives.
- They often use the *National Security Act* to justify their destructive deeds and to provide cover when they get caught.

So who are "they?" Apparently, Wall Street, multinational corporations, international banks and wealthy people have formed elite groups to serve as vehicles for their interests. These include *Think Tanks* such as the Council on Foreign Relations, the Trilateral Commission, the Bilderbergers and others, which are interlocked with the Tax-exempt Foundations and the Federal Reserve. According to Professor Sutton, Think Tanks, the Federal Reserve, the Executive Branch of the White House, Law, Education, and Media are all interlocked at the top, and controlled by an elite cabal.

"We should realize that many of these groups are closely allied together," agreed Professor Marrs. "In some cases," he said, "the same man may simultaneously serve as either president or on the board of directors of a banking and money institution, a political group, an intelligence organization, a religious denomination or organization, and a foundation or educational group." Perloff acknowledged this interlock too, when he stated that their members serve "on the boards of vast foundations," such as "Rockefeller, Ford, and Carnegie."

Rene Wormser, General Council to the Reece Committee which investigated Tax-exempt Foundations, provides us with an example of interlocking directorates. In his book, *Foundations: Their Power And Influence*, he sites a report from a congressional investigation by the Reece Committee, concerning the Rand Corporation. He states, "This is a corporation in the nature of a foundation, which plays a very important part in government research. It would warrant special attention in connection with any study of the extent to which foundation interlocks have influenced government."

"Among the trustees and officers of The Rand Corporation," continues Wormser, "were found [by the Reece Committee] the following who had material connections with other foundations: Charles Dollard (trustee), Carnegie Corporation; L.A. Dudbridge (trustee), Carnegie Endowment National Science Foundation; H. Rowan Gaither, Jr. (trustee) Ford Foundation; Philip E. Mosely (trustee), Ford Foundation ... Frederick F. Stephan (trustee), Rockefeller Foundation ... Hans Speier (officer) (Ford) Behavioral Science Division..."

These Think Tanks hold regular meetings, which are not open to the public. According to a book entitled, *Who's Who of the Elite*, by Robert Gaylon Ross, topics of conversation include wars. He explained, "[they] decide when wars should start, how long they should last, when they should end, who will and will not participate, the changes in boundaries of countries resulting from the outcome of these wars, who will lend the money to support the war efforts, and who will lend the money to rebuild the countries after they have been destroyed by war."

"They own the central banks," continued Ross, "such as the Federal Reserve System in the US, and similar organizations in all major countries throughout the world, and therefore are in a position to determine discount rates, prime rates ... [and] what countries should receive loans (guaranteed by the taxpayers of the respective countries). ... They decide who will be allowed to run for the offices of President, Prime Minister, Chancellor, Governor General, or other names applied to the leaders of all major countries around the world." In addition, they, "directly or indirectly own all the major news media, and can therefore tell the public exactly what they want them to hear, and deny the public information they do not want them to see, hear, or read."

If the general population knew that a handful of unelected people were running their government they would probably be outraged. This "legitimate mafia" prefers to remain behind the scenes, apparently so people will still believe they live in a constitutional republic. But their existence is not completely hidden. By incident, or accident, people such as targeted individuals, corporate/government whistleblowers and others, have directly experienced how government really works and who controls it. Their experience on the front lines have given them insights into approaching situations that the general population is not yet aware of.

"Secret elitist groups always censor, or try to censor, news about their covert activities," declared Antony Sutton and Patrick Wood in their book, *Trilaterals Over Washington*. "Censorship stems from the overall need for secrecy, to conceal from the world at large. As long as Trilaterals (and Bilderbergers and other elite groups) skulk around the world convening closed meetings in secluded corners with security guards to keep out the press then we may conclude that Trilaterals, Bilderbergers, and the rest have something to hide."

Most members of these elite organizations are not aware of their true objectives. According to these researchers, there is an inner-core of members that are. "These members are aware of only about 50%, or less of the goals, and objectives of the Global Union movement," proclaimed Ross. "A large number of these people are members for ego, and social reasons only," and would probably resign if they were to "find out what the Global Union is 'really' up to."

These elite organizations have been called Think Tanks or modern secret societies.(**) Some receive their funding from Tax-exempt Foundations. "The existence of groups like the Trilateral Commission, Council on Foreign Relations, and the Bilderbergers is well documented," stated Jim Marrs in his underground bestseller, *Rule by Secrecy*. "The only question is the extent of their control and manipulation of major world events."

"Likewise," said Marrs, "there is no question that members of these societies exert inordinate control over many of the largest corporations and banks in the world. These corporations, in turn, control essential minerals, energy, transportation, pharmaceuticals, agriculture, telecommunications, and entertainment—in other words, the basics of modern life."

When these policies are set at the top by the controlling elite who have been known to promote a one-world socialist order, then it's understandable how destructive policies, which appear to be positive, can be filtered down to local and state governments. Marrs explained, "If the top leadership of government and business is controlled ... then the activities of subservient agencies and divisions must be of little consequence. Government bureaucrats—honest and well-intentioned workers for the most part—simply follow orders and policies set by superiors."

Former FBI Agent Dan Smoot acknowledged this covert power structure in his book, *The Invisible Government*, when he wrote, "Somewhere at the top of the pyramid in the invisible government are a few sinister people who know exactly what they are doing..." The invisible government that Smoot refers to has established identifiable groups, which it uses as vehicles to carry out its policies. In his bestselling book, *None Dare Call It Conspiracy*, Author Gary Allen, declared that they have "established some very special and highly influential organizations, many of which have clandestine aims and goals." He lists some of these groups as, "the Bilderbergers, the Club of Rome, the Aspen Institute, the Trilateral Commission, the Council on Foreign Relations, [and] the Bohemian Grove."

Local, state, and federal law enforcement are reportedly controlled by these Think Tanks. "The Elite control our courts, the Pentagon, the U.N., NATO, N.S.A., C.I.A., F.B.I., B.A.T.F., our Senate and House of Representatives," charged Ross, "and directly or indirectly control all local law enforcement agencies."

Author Marrs agreed that government agencies under control of these elite groups include, "not only the Central Intelligence Agency (CIA)," he said "but the National Security Council (NSC), Federal Bureau of Investigation (FBI), National Security Agency (NSA), Defense Intelligence Agency (DIA), National Reconnaissance Office (NRO), Drug Enforcement Agency (DEA), Bureau of Alcohol, Tobacco and Firearms (BATF), Internal Revenue Service (IRS), Federal Emergency Management Agency (FEMA), and many others." He added, "These agencies are themselves secretive, citing reasons of national security, [and] executive privilege."

"The Rockefellers, Rothschilds, and a few others—numbering less than a dozen leaders of international finance are the real power behind the visible thrones of world government," observed Professor Marrs. "No major policy is formulated without their input; no major plan of action is implemented without their specific 'go' signal."

One of my goals is to show you who "they" are, what they've done, how they've covered it up, and what their plans are. Throughout the rest of this book, other authors and I will periodically refer to these interlocking Think Tanks using the abbreviations, *BB* for Bilderberg, *TC* for Trilateral Commission, and *CFR* for Council on Foreign Relations. Now I'll explain these elite groups in more detail. Let's meet your unelected rulers.

Endnotes

* From the information I've come across, the term *New World Order* has been used to describe three things: a political movement, a group of individuals bringing about the movement, and joint reference to both the movement and the group fostering it. My understanding is that the term refers to the movement.

** Some literature describes these organizations as think tanks or modern secret societies, and others describe them as foundations. Carnegie Endowment is listed as a foundation in *Foundations: Their Power and Influence* by Wormser, but also listed as a think tank in *Who's Who of the Elite* by Ross, and *Trilaterals Over Washington* by Sutton and Wood. Author Marrs describes the CFR, TC and Bilderbergers as modern secret societies.

Chapter 3

The Trilateral Commission

The Trilateral Commission was founded in 1973 by David Rockefeller and Zbigniew Brzezinski. It has headquarters in New York, Paris, and Tokyo. It is said to be funded by the tax-exempt giants like Ford, Lilly Endowment, Rockefeller Brothers Fund, the German Marshall Fund, and corporations such as Time, Bechtel, Exxon, General Motors, Wells-Fargo, etc.

Its membership is composed of past and present presidents, ambassadors, secretaries of state, Wall Street investors, NATO and pentagon military personnel, international bankers, foundation executives, media owners, university presidents and professors, senators and congressmen, and wealthy industrialists and entrepreneurs.

"The Trilateral Commission was formally established in 1973 and consisted of leaders in business, banking, government, and mass media from North America, Western Europe, and Japan," wrote James Perloff. "David Rockefeller was founding chairman and Brzezinski founding director of the North American branch, most of whose members were also in the CFR," he added. Author Jim Marrs announced, "With the blessing of the Bilderbergers and the CFR, the Trilateral Commission began organizing on July 23-24, 1973 at the 3,500-acre Rockefeller estate at Pocantico Hills, a subdivision of Tarrytown, New York."

Its existence is a proven fact as verified by mainstream news sources such as *Time Magazine*, the *Wall Street Journal*, *U.S. News and World Report*, and *The Washington Post*, as well as multiple books including, *Trilaterals Over Washington*, by Professor Antony Sutton and Patrick M. Wood, *America's Secret Establishment*, by Professor Antony Sutton, *Dark Majesty* by Professor Texe Marrs (no known relation to Jim Marrs), *The Shadows of Power*, by James Perloff, and others.

Professor Antony Sutton wrote, "The organization is completely above ground. In fact," he stated, "this author has openly debated with George Franklin, Jr., ... of the Trilateral Commission on the radio. Mr. Franklin did show a rather ill-concealed dislike of the assault on his pet global New World Order—and made the mistake of attempting to disguise this objective."

The Trilateral Commission has an official publication called, *Trialogue*. It also issues multiple *Task Force Reports* (also called *Triangle Papers*) per year. Although its meetings are invitation-only, its membership list is publicly available and can be obtained by contacting them at: 345 East 46th Street, Suite 711, New York, NY 10017.

Authors Sutton and Wood comment about the seemingly innocent objectives of the Trilateral Commission. Quoting from the objectives which appear in every issue of *Trialogue*, they wrote, "The Trilateral Commission was formed in 1973 by private citizens of Western Europe, Japan and North America to foster closer cooperation among these three regions on common problems. It seeks to improve public understanding of such problems, to support proposals for handling them jointly, and to nurture habits and practices of working together among these regions." Sutton and Wood then explain that the rest of their book is devoted to telling the *truth* about the commission.

Referring to the *Task Force Report* entitled, *The Crisis of Democracy* written in part by Harvard political scientist Samuel P. Huntington, author Marrs noted, "The paper suggested that leaders with 'expertise, seniority, experience and special talents' were needed to 'override the claims of democracy.'" He added, "Three years after his paper was published, Huntington was named coordinator of the 1979 presidential order creating the Federal Emergency Management Agency [FEMA], a civilian organization with the power to take totalitarian control of government functions in the event of a national 'emergency.'"

Like the Council on Foreign Relations, the Trilateral Commission has infiltrated the executive branch of the United States government. Regarding the beginning of this pattern of infiltration into the White House, Sutton and Wood commented, "On 7 January 1977 *Time magazine*, whose editor-in-chief, Hedley Donovan, is a powerful Trilateral Commissioner, named President Carter 'Man of the year.'" They added, "Carter had already chosen his cabinet. Three of his cabinet members—Vance, Blumenthal, and Brown—were Trilateral Commissioners... In addition, Carter had appointed another fourteen Trilateral Commissioners to top government posts. ... These presidential appointees represented almost one-third of the Trilateral Commission members from the United States. Try to give odds to that!"

The January 16, 1977 issue of *The Washington Post* expressed, "Trilateralists are not three-sided people. They are members of a private, though not secret, international organization put together by the wealthy banker, David Rockefeller, to stimulate the establishment dialog between Western Europe, Japan and the United States. But here is the unsettling thing about the Trilateral Commission. The President-elect [Carter] is a member. So is the Vice-President-elect Walter F. Mondale. So are the new Secretaries of State, Defense and Treasury. So is Zbigniew Brzezinski, who is a former Trilateral Director and Carter's National Security Adviser, also a bunch of others who will make foreign policy for America in the next four years."

The same issue of *The Washington Post* reported that, "At last count, 13 Trilateralists had gone into top positions in the administration, not to mention six other Trilateralists who are established as policy advisers, some of whom may also get jobs. This is extraordinary when you consider that the Trilateral Commission only has about 65 American members."

Author Perloff wrote, "The new President [Carter] appointed more than seventy men from the CFR, and over twenty members of the much smaller Trilateral Commission. Zbigniew Brzezinski acknowledges in his *White House Memoirs*: "Moreover, all the key foreign policy decision makers of the Carter Administration had previously served in the Trilateral Commission..." "Brzezinski," stated Perloff, "of course, became National Security Adviser, the same position Kissinger had held."

Commenting on a June 18, 1974 article in the *New York Times*, which stated, "the lives and fortunes of large numbers of human beings hang upon the outcome of decisions taken by a small handful of national leaders—on the Trilateral Commission," Gary Allen warned that "it was time to pay more attention—a lot more attention—to the group."

Regarding the consolidation process of the New World Order, Sutton and Wood commented, "In September 1974 Brzezinski was asked in an interview by the Brazilian newspaper *Vega*, 'How would you define this new world order?'" Brzezinski answered, "We need to change the international system for a global system in which new, active and creative forces—recently developed—should be integrated. This system needs to include Japan, Brazil, the oil producing countries, and even the USSR..." When asked if Congress would have an expanded or diminished role in the new system, Brzezinski declared, "The reality of our times is that a modern society such as the U.S. needs a central coordinating and renovating organ which cannot be made up of six hundred people."

This man is telling you that this New World Order will be a consolidation of individual countries into a single world government, which will be controlled by a small clique of insiders, who believe they are fit to rule the planet. Anytime you have a consolidation of the power such as this, you have a dictatorship.

Quoting from Brzezinski's book, *Between Two Ages*, Sutton and Wood wrote that Brzezinski described Marxism as, “a further vital and creative stage in the maturing of man's universal vision. ... Tension is unavoidable as man strives to assimilate the new into the framework of the old. But at some point the old framework becomes overloaded.” Brzezinski continued, “The new input can no longer be redefined into traditional forms, and eventually it asserts itself with compelling force. Today, though the old framework of international politics—with ... the fiction of sovereignty ... is clearly no longer compatible with reality.”

According to Sutton and Wood, when Brzezinski uses the word “framework” he apparently means the U.S. Constitution; if this is so, then what he told us in 1971 is that the Trilateral Commission plans to make drastic changes to the U.S. Constitution. “One of the most important ‘frameworks’ in the world ... is the United States Constitution,” exclaimed Sutton and Wood. So why is it so important to these elite organizations that the Constitution be changed?

Well, they’ve admitted in their own publications that they tend to merge the U.S. and other NATO countries into a single world government controlled by the big corporations. Would a constitution which *guarantees* individual freedom interfere with their plan? Before the amendments made by the U.S. Patriot Act, and other anti-terror legislation, the constitution made it impossible for a totalitarian regime to flourish in the United States.

Sutton and Wood wrote that the Trilaterals wanted to assemble, “a national constitutional convention to re-examine the nation's formal institutional framework,” in order to open up “a national dialog on the relevance of existing arrangements... The needed change,” said the Trilaterals “is more likely to develop incrementally and less overtly.” Brzezinski himself declared, “International banks and multinational corporations are acting and planning in terms that are far in advance of the political concepts of the nation-state.”

Changing “existing arrangements” can be safely translated to *existing freedoms*, such the Bill of Rights. And “less overtly” means they'll be making these changes without your approval using deception. Sutton and Wood commented, “When Brzezinski refers to “develop(ing) incrementally and less overtly” he is specifically recommending a deceptive ... approach to abandonment of the Constitution." After the current "framework" is removed, it will apparently be replaced with a world constitution furnished by the UN, which doesn't *guarantee* personal freedom as a human right, but makes it a *privilege*, which is granted if possible.

When multinational corporations and banks run the planet, this is basically global fascism. Whether it's called fascism, or communism, or socialism, it's all the same, which is control by those in charge of the state, or super-state, in the case of the New World Order. Again Sutton and Wood warn, "Those ideals which led to the heinous abuses of Hitler, Lenin, Stalin, and Mussolini are now being accepted as necessary inevitabilities by our elected and appointed leaders."

"Fascism should more appropriately be called corporatism because it is a merger of state and corporate power."
-Benito Mussolini

In his bestselling book, *With No Apologies*, Senator Barry Goldwater described the true intentions of the Trilateral Commission as, "a skillful, coordinated effort to seize control and consolidate the four centers of power—political, monetary, intellectual, and ecclesiastical [religious]." He added, "All this is to be done in the interest of creating a more peaceful, more productive world." In other words by using a *big lie*.

The *Task Force* report put out by the TC entitled, *The Crisis of Democracy* stated, "The democratic political system no longer has any purpose. The concepts of equality and individualism give problems to authority. The media is not sufficiently subservient to the elite. Democracy has to be "balanced" (i.e., restricted). The authority and power of the central government must be increased."

These people are telling you that they're going to restrict your individual rights and centralize power into corporate hands. There can be no confusion over the objectives of this group that has infiltrated the executive branch since the 1970s. They are telling you in their own publications exactly what they intend to do. *They are setting up a worldwide fascist dictatorship*!

"Trilateralism is the current operational vehicle for a corporate socialist takeover," advised Sutton and Wood. Likewise, Senator Goldwater calls the Trilateral Commission an "international cabal," which "is intended to be the vehicle for multinational consolidation of the commercial and banking interests by seizing control of the political government of the United States." This covert takeover has been done, not by a civil war but by infiltration.

On July 23, 1976, the Greek newspaper *Exormisis* recognized this overthrow when they wrote, "A new kind of fascism emerges with Carter. The oppression will not have the form we used to know, but it will be the 'depoliticization' of all citizens in the U.S., and the generating of all power in the executive branch, that is, the Presidency, without the President giving any account to the Congress or anybody else except the multinationals [Banks/Corporations], which have financed Carter's campaign... The accession to power of Carter ... would mean a new era of dictatorial policies."

"Like sheep going to slaughter, our people cannot smell the death that awaits them," warned Sutton and Wood. "If we are about to be thrown into the pits of the dark ages, the most logical catalyst, or motivator on the horizon is the TRILATERAL COMMISSION." [Emphasis in original]

Summary

The Trilateral Commission presents itself as a humanitarian group, which seeks to promote world peace through understanding and cooperation. However, its publications indicate that it plans to install a worldwide dictatorship, which will be dominated by the multinational corporations. This is being accomplished using deception.

Chapter 4

The Council on Foreign Relations

Congressman James E. Jeffries wrote, "If the Establishment is elusive in its identity, it certainly has a perceptible face in the Council on Foreign Relations..." Like its counterpart The Trilateral Commission, the CFR is composed of Wall Street investors, international bankers, foundation executives, members of Think Tanks and Tax-exempt Foundations, ambassadors, past and present presidents, secretaries of state, lobbyist lawyers, media owners, university presidents and professors, federal and supreme court judges, and members of military leaders from NATO and the pentagon.

It was formally established in New York, on July 29, 1921, as a counterpart to a British group called, The Royal Institute of International Affairs (RIIA). "The CFR and RIIA were originally intended to be affiliates, but became independent bodies, although they have always maintained close informal ties," wrote James Perloff, in his well-documented book, *The Shadows of Power*.

Its meetings are not open to the public and membership is by invitation only. Like the Trilateral Commission, its membership list is publicly available. The CFR puts out a publication known as *Foreign Affairs* which has been called "the most influential periodical in print," by *Time Magazine*. They are located at the Harold Pratt House, 58 East 68th Street in New York City, New York 10021.

It was originally dominated by J.P. Morgan interests, and began to gain momentum around 1927 with funding from the Rockefeller and Carnegie Foundations. The founding president was John W. Davis, millionaire and personal attorney of J. P. Morgan. Some other early/founding CFR members included, Henry Davison, Thomas Thacher, Harold Swift, W. Averill Harriman, John Foster Dulles, Allen Dulles, Thomas Lamont, Paul Cravath, Federal Reserve architect Paul Warburg, Mortimer Schiff (Jacob's son), Morgan partner Russell Leffingwell, and other Morgan partners.

Jim Marrs wrote in his book *Rule by Secrecy* that, "Funding for the CFR came from bankers and financiers such as Morgan, John D. Rockefeller, Bernard Baruch, Jacob Schiff, Otto Kahn, and Paul Warburg. Today, funding for the CFR comes from major corporations such as Xerox, General Motors, Bristol-Meyers Squibb, Texaco, and others as well as the German Marshall Fund, McKnight Foundation, Dillion Fund, Ford Foundation, Andrew W. Mellon Foundation,

Rockefeller Brothers Fund, Starr Foundation, and the Pew Charitable Trusts." The CFR created junior chapters in most major cities called, *Committees on Foreign Relations*. These junior chapters hold periodic dinner meetings.

Many researchers on this subject agree that there is an inner and outer core of the CFR and its interlocking Think Tanks. Members of the outer ring are just window dressing or camouflage, and are not aware of the motivating factors of the inner-core. The inner-core appears to be members of the steering and advisory committees.

Similar to the TC objectives, those of the CFR seem humanitarian on the surface. Describing this façade in his book, *Who's Who of the Elite*, Robert Gaylon Ross explained, "Let's start with the smoke and mirrors furnished by the CFR in several of their Annual Reports." Quoting a CFR Annual Report from 1993-4, he said, "The Council on Foreign Relations is a non-profit, and non-partisan membership organization dedicated to improving the understanding of U.S. foreign policy, and international affairs through the exchange of ideas." However, he states, "if you are doing something illegal, immoral, unethical, unpopular, and/or unconstitutional, you will do whatever is necessary to see that it is kept secret."

Professor Sutton describes the CFR as "superficially an innocent forum for academics, businessmen, and politicians, [which] contains within its shell, perhaps unknown to many of its members, a power center that unilaterally determines U.S. foreign policy." He states that their true "subversive" objective is "the acquisition of markets and economic power ... for a small group of giant multinationals under the virtual control of a few banking investment houses and controlling families."

Former FBI Agent Dan Smoot observed the same veil of deception. In his book, *The Invisible Government*, he wrote, "The leadership of the invisible government doubtless rests in the hands of a sinister ... few." Regarding the majority of members he stated, "Many, if not most, of these are status-seekers." But warned, "The ultimate aim" of the CFR, "however, well-intentioned its prominent and powerful members may be" is "to create a one-world socialist system." When you see the word *socialism* think *tyrannical dictatorship*, dominated by the big corporations and international banks.

"A number of individuals are apparently invited into the CFR simply because they have a distinguished name or other enhancing qualities," stated Perloff. He continued, "[they] join without endorsing or even knowing the Council's habitual viewpoint." "However," he said, "The membership's great majority ... have been chronically pro-socialist and pro-globalist." Professor Sutton recognized, "most members of the CFR have no knowledge of this diabolical plan. But there is an inner core within the CFR that ... promotes it."

Regarding the media blackout, Allen observed, "During its first fifty years of existence, the CFR was almost never mentioned by any of the moguls of the mass media." "And," he added, "When you realize that the membership of the CFR includes top executives from the *New York Times*, *The Washington Post*, the *Los Angeles Times*, the *Knight Newspaper* chain, *NBC*, *CBS*, *Time*, *Life*, *Fortune*, *Business Week*, *US News and World Report*, and many others, you can be sure that such anonymity is not accidental."

Referring to the whitewashed Reece Committee investigations which feebly investigated the Tax-exempt Foundations interlocked with the CFR, Smoot wrote, "The power of the Council is somewhat indicated by the fact that no committee of Congress has yet been powerful enough to investigate it or the foundations with which it has interlocking connections and from which it receives its support." He declared, "In 1939, the Council began taking over the U.S. State Department."

Admiral Chester Ward, former Judge Advocate General of the U.S. Navy, remained in the CFR for about 20 years and co-authored a book entitled, *Kissinger on the Couch*, where he wrote, "Once the ruling members of the CFR have decided that the U.S. Government should adopt a particular policy, the very substantial research facilities of the CFR are put to work to develop arguments, intellectual and emotional, to support the new policy, and to confound and discredit, intellectually and politically, any opposition."

The Admiral also warned that the goal of the CFR is the "submergence of US sovereignty and national independence into an all-powerful one-world government." In the August 1978 issue of *W Magazine*, former CFR President Winston Lord is quoted as saying, "The Trilateral Commission doesn't secretly run the world. The Council on Foreign Relations does that."

Rene Wormser who served on the Reece Committee investigating the multi-billion-dollar Tax-exempt Foundations and their interlocks, wrote in his book, *Foundations*, that "The Council on Foreign Relations" is "virtually an agency of the government," and that it is "financed both by the Rockefeller and Carnegie foundations, [and furthermore it] overwhelmingly propagandizes the globalist concept."

Author Allen agreed when he wrote, "The C.F.R. is totally interlocked with the major foundations and [other] so-called 'Think Tanks.'"

Regarding the interlocking Think Tanks, Dan Smoot explained, "All of the organizations have federal tax-exemption as 'educational' groups; and they are all financed, in part, by tax-exempt foundations, the principal ones being Ford, Rockefeller, and Carnegie. Most of them also have close working relations with official agencies of the United States Government." Referring to the CFR's infiltration into the White House, Senator Goldwater wrote that the CFR has "staffed almost every key position of every administration since that of FDR."

The senate was apparently concerned about the influence and infiltration of the CFR into the White House. Quoting a *Congressional Record*, dated December 15, 1987, vol. 133, Perloff wrote, "Senator Jesse Helms, after noting the CFR's place within the Establishment, put it this way before the Senate in December 1987: The viewpoint of the Establishment today is called globalism. ... Mr. President, in the globalist point of view, nation-states and national boundaries do not count for anything. Political philosophies and political principles seem to become simply relative. Indeed, even constitutions are irrelevant to the exercise of power..."

In an article called, *School for Statesmen*, which appeared in the July 1958 issue of *Harpers*, CFR member Columnist Joseph Kraft, proclaimed, "It [the CFR] has been the seat of some basic government decisions, has set the context for many more, and has repeatedly served as a recruiting ground for ranking officials." Allen commented, "The policies promoted by the C.F.R. in the fields of defense and international relations become, with regularity which defies the laws of chance, the official policies of the United States Government."

"Today the C.F.R remains active in working toward its final goal of a government over all of the world—a government which the Insiders and their allies will control," declared Allen. "The goal of the C.F.R. is simply to abolish the United States with its constitutional guarantees of liberty. And they don't even try to hide it." Study No. 7, published by the CFR on November 25, 1959, advocates "building a new international order [which] must be responsive to world aspirations for peace, [and] for social and economic change." This new order will include, "states labeling themselves as 'Socialist' [Communist]."

Allen refers to the CFR as the "invisible government," and says it is "unquestionably" the "most influential group in America." Former Agent Smoot concurs, writing, "I am convinced that the Council on Foreign Relations, together with a great number of other associated tax-exempt organizations, constitutes the invisible government which sets the major policies of the federal government."

Continuing, he said, the CFR, "exercises controlling influence of government officials who implement the policies; and, through massive and skillful propaganda, influences Congress and the public to support the policies." He stated further that, "the objective of this invisible government is to convert America into a socialist state and then make it a unit in a one-world socialist system."

"We shall have world government whether or not you like it—by conquest or consent."
-CFR member James Warburg, testifying at Senate Foreign Relations Committee on February 17, 1950

Regarding the CFR's plan for a single world government, Perloff said, "The CFR advocates the creation of a world government." And, "Anyone who cares to examine back issues of *Foreign Affairs* will have no difficulty finding hundreds of articles that pushed ... this concept of globalism."

"For decades," said Perloff, "the CFR pushed this ascending approach to world government, with *Foreign Affairs* carrying such titles as *Toward European Integration: Beginnings in Agriculture*, *Toward Unity in Africa*, *Toward a Caribbean Federation*, and so on." He quotes a 1974 article which stated, "The house of world order will have to be built from the bottom up rather than from the top down ... an end run around national sovereignty, eroding it piece by piece, will accomplish much more than the old-fashioned frontal assault."

Very few people are aware of the existence of the CFR or its purpose. This appears to be the case with their interlocking Think Tanks such as the Bilderbergers and the Trilateral Commission. Perloff said, "Today, probably not one American in five hundred can identify the CFR, despite the fact that it is arguably the most powerful political entity in the United States." And Allen's estimate is higher; he comments, "It is doubtful that one American in a thousand so much as recognizes the Council's name, or that one in ten thousand can relate anything at all about its structure or purpose."

Summary

The CFR is a supra-governmental organization that overshadows congress. It is financed by the Tax-exempt Foundations, and portrays itself as a humanitarian group. But its objectives are to erode national sovereignty and merge America into a worldwide government under the control of its members, which include people of tremendous wealth. Basically, their goal is *global fascism.* The CFR is responsible for setting major policy, which is activated without public knowledge or consent. It is beyond congressional investigation.

Chapter 5

The Bilderbergers

The Bilderbergers are an international group composed of European Royalty, Wall Street investors, politicians, international bankers, prominent businessmen, media executives, and military leaders from around the planet. "Those in attendance include leading political and financial figures from the United States and Western Europe," proclaimed Allen. Allen also says that it is probably "not accidental," that less than one person in 5,000 is aware of this group. Jim Marrs wrote, "The Bilderbergers are a group of powerful men and woman—many of them European royalty—who meet in secret each year to discuss the issues of the day."

The first official meeting of the Bilderbergers took place at the Hotel de Bilderberg in Oosterbeek, Holland in May 1954. It was formed in part by Prince Bernhard of the Netherlands, a former Nazi. "The primary impetus for the Bilderberger meetings came from Dutch Prince Bernhard," wrote author Marrs, "a former member of the Nazi Schutzstaffel (SS) and an employee of Germany's I.G. Farben in Paris." The Rockefellers and Rothschilds are said to have played an important part in its inception also.

Veteran journalist James P. Tucker has investigated the Bilderbergers nearly 30 years. His book, *The Bilderberg Diary*, which includes photographic evidence, is a well-documented account of the results of his investigation. In it he writes, "The Bilderberg group is an organization of political leaders and international financers that meets secretly every spring to make global policy. There are about 110 regulars," he says, including, "Rockefellers, Rothschilds, bankers, heads of international corporations and high government officials from Europe and North America." These people meet at varying locations around the planet.

Referring to the financial elite's various fronts, Professor Marrs wrote, "Comprising themselves in the form of the Bilderberger Group, they have met at least annually since 1954." Their meetings are by invitation only, and according to Gaylon Ross, of the three interlocking Think Tanks, they are the most secretive. They are also the most influential. Their meetings are financed by the Tax-exempt Foundations. Allen asked, "Why do the Ford, Rockefeller and Carnegie foundations finance the meetings if they are not important?"

In 1964, Dan Smoot acknowledged this elite group when he wrote, “The ‘Bilderbergers’ are another powerful group involved in the internationalist web.” Speaking of the interlocking membership, he proclaimed, “American firms associated with the society are said to be among the large corporations whose officers are members of the Council on Foreign Relations and related organizations.”

The Bilderbergers have close ties to Europe’s nobility such as the British royal family, as well as royalty from Holland, Sweden and Spain. They also contain the same representatives in finance and industry as their counterparts, The Trilateral Commission and the Council on Foreign Relations. There is also a heavy cross-membership with the Trilateral Commission and the Council on Foreign Relations.

“As with the Trilateral Commission and the Council on Foreign Relations, Bilderbergers often carry cross-membership in two or more of these three groups, commented Marrs.” Allen remarked, “Those who adhere to the accidental theory of history will claim that it is sheer coincidence that every single one of those named as past and present members of the Bilderberger Steering Committee is or was a member of the Council on Foreign Relations.”

The Bilderbergers are said to be the organization that sets global policy which is then passed on to the CFR and TC to implement. While the Trilateral Commission and the Council on Foreign Relations both publish their membership rolls, the Bilderbergers keep the meetings, goals, and membership lists secret. “Bilderberg does not publish membership lists but they have been obtained by investigative reporters working for *The Spotlight*,” wrote Ross.

Each year the hosting government is responsible for providing heavy security for three days, which includes: the military, the secret service, federal and local police, and private security firms. Allen explains, “The participants are housed in one location and are protected by a thorough security network. Decisions are reached, resolutions adopted, plans of action initiated...”

“When they meet, they clear out all the guests, and employees in the building in which they are to meet,” Ross said, and added “they completely de-bug all the rooms, bring in their own cooks, waiters, housekeepers, heavily armed security guards, etc. and do not allow ‘outsiders’ anywhere near the meeting place just before, during and immediately after they meet.”

Continuing he explained, “Each time they have met on US soil, the meetings were held on Rockefeller owned property.” According to veteran journalist Tucker, some hotel staff is allowed to stay but are given strict orders to never to speak to a Bilderberger unless spoken to, or look them in the eyes.

The *Rutland Vermont Herald* apparently received notice of a Bilderberg meeting when local officials were given instructions on how to prepare for it. It declared on April 20, 1971, "A rather tight lid of secrecy was being kept on the [Bilderberg] conference." "A closed-door meeting," it described, "was held in Woodstock last week to brief a handful of local officials on some phases of the [future] conference. ... The Woodstock Inn will apparently be sealed up like Fort Knox. ... No press coverage will be allowed."

Referring to the same meeting which took place on April 23, 1971 in Woodstock, Vermont, Allen wrote, "When Prince Bernhard arrived at Boston's Logan Airport, he did admit to reporters that the subject of the conference would be the "change in the world-role of the United States." Allen remarked, "Isn't it nice to have changes in America's role in the world decided upon by Bernhard, Rothschild and Rockefeller?"

The mainstream news will normally not report on these meetings. Speaking of the media blackout Tucker said, "Bilderberg has, at one time or another, had representatives of all major U.S. newspapers and network news outlets attend. They do so on their promise to report nothing. This is how Bilderberg keeps its news blackout virtually complete in the United States." According to Tucker, as part of a disinformation strategy, mainstream newspapers in the United States may even portray these meetings as *harmless talk-sessions* of little importance.

Marrs commented, "Unlike their American counterparts, some members of Scotland's news media found their voice. Under the headline, *Whole World in their Hands*, Jim McBeth of *The Scotsman* described the tight security surrounding the [1998] meeting [at the Turnberry Hotel near Glasgow, Scotland], commenting, 'Anyone approaching the hotel who did not have a stake in controlling the planet was turned back.'"

"The press, naturally, is not allowed to be present, although occasionally a brief press conference is held at the end of the meeting at which time the news media are given in very general terms the Bilderberger version of what was discussed," observed Allen. Reporters that are not invited and attempt to document these meetings have been arrested. Marrs added, "Despite the fact that many highly regarded American media members meet with the Bilderbergers, little or nothing gets reported on the group or its activities..."

Although there is a media blackout concerning these meetings, the news media are always present, agreed Ross. He lists past attendees, "such as: Peter Jennings (BB, and Anchor and Senior Editor of *ABC News*, *World News Tonight*), Joseph C. Harsch (BB, CFR, and former Commentator for *NBC, Inc.*), Bill D. Moyers (BB, and Executive Director of Public Affairs TV, Inc., and former Director of the CFR), [and] William F. Buckley, Jr. (BB, CFR, and Editor-in-Chief of *National Review*, and host of *PBS's Firing Line*)."

Ross says other media personnel which have attended include, "Gerald Piel (BB, CFR, and former Chairman of Scientific America, Inc.), Henry Anatole Grunwald (BB, CFR, and former Editor-in-Chief of *Time, Inc.*), Mortimer B. Zuckerman (BB, CFR, and Chairman and Editor-in-Chief of the *US News and World Report*, *New York Daily News*, and *Atlantic Monthly*), Robert L. Bartley (BB CFR, TC, and Vice President of the *Wall Street Journal*)."

More media attendees, according to Ross, are Peter Robert Kann (BB, CFR) of Dow Jones and Company, Donald C. Cook of the *Los Angeles Times*, Thomas L. Friedman of the *New York Times*, and Katherine Graham owner of *The Washington Post*. Most of these attendees have cross memberships in the CFR and TC.

Tucker commented that mainstream media apparently doesn't think these meetings are newsworthy. Evidently, for decades the *Post* and most other major media outlets didn't think the public would be interested in learning about these world leaders meeting to shape world events. Is it possible that there's something else of major significance they're not reporting or are being used to cover-up?

On May 26, 1995 the Swiss newspaper, *Tagesschau*, ran a story entitled, *The Swiss Military to Protect International and Leading Economists and Statesmen.* The article explained, "Prominent leaders from the world of politics and economics will be meeting at Nidwalden on Mt. Bergenstock, Switzerland, for the very secretive Bilderberg conference. ... [A]mong the participants are Leading Bankers, politicians and industrialists from Europe and North America. The Luminaries from the 1995 meeting include Kurt Furgler, Walter Scheel, Henry Kissinger, Helmut Kohl, Fiat Chairman Giovanni Agnelli, shipping magnate Stavos Niarchos, David Rockefeller and Prince Claus of the Netherlands."

The 1994 meeting in Helsinki, Finland, Tucker wrote, included, "In addition to Rockefeller and Kissinger, other familiar faces ... [such as] Lord Peter Carrington, Queen Beatrix of Holland, NATO chief Woerner, President Ahtisaari and Prime Minister Esko Aho of Finland,

German Chancellor Helmut Kohl, German central banker Hans-Otto Pohl and Atos Erkko, a well known Finnish publisher."

Also in attendance says Tucker, were, "Franz Vranitsky, President of Austria; Percy Barnevik, President of ABB (Asea Brown Boveri Ltd.) of Sweden; Giovanni Agnelli, head of the giant Fiat firm in Italy; Max Jacobson and Jaakko Illoniemi of Finland; Rozanne Ridgway, Assistant Director of the White House Office of Management and Budget; and Volker Ruhe of Germany." Tucker lists other attendees as, "Katherine Gram, owner of *The Washington Post Co.*; Louis Gerstner of IBM; Thomas Pickering, U.S. Ambassador to Russia; Brent Scowcroft, Adviser to Bush; Paul Allaire of Xerox; Peter Sutherland of Britain; [and] Queen Sofia of Spain."

Referring to the meeting in Versailles, France in 2003, Tucker announced, "The following have been positively identified as participating: Queen Beatrix of The Netherlands; Ali Babacan, Minister of the Economy in Turkey; King Juan Carlos and Queen Sophia of Spain; Jacques Chirac, President of France, Kenneth Clark, former British Chancellor of the Exchequer and member of the Parliament; Etienne D'avignon, Societe General of Belgium; Jean Louis Debre, President of the French National Assembly; Kermal Dervis, Turkey; Sevein Gjerem, CEO of the National Bank of Norway."

Gerald Ford attended Bilderberg meetings 1964 and 1966, and in 1974 became President. And while Governor of Arkansas, Bill Clinton attended a 1991 meeting before becoming President in 1992. Regarding Clinton's Bilderberg connection, Marrs wrote, "In 1991, then Arkansas Governor Bill Clinton was honored as a Bilderberg guest. The next year he ran for and won the presidency of United States."

The plan for a common European currency, the *Euro*, is said to have been formulated by the Bilderbergers. Tucker claims that the plan for a "European super-state" and the "euro" had been on the Bilderberg agenda for years. "It was viewed as a major step toward their goal of a world government and creates a favorable climate for the huge banks to consume the small ones and for the huge, international conglomerates to absorb the small firms," he observed.

Quoting a Bilderberger, Marrs added that, "Jack Sheinkman, Chairman of Amalgamated Bank and a Bilderberger member, stated in 1996, 'In some cases discussions do have an impact and become policy. The idea of a common European currency was discussed several years back before it became policy. We had a discussion about the U.S. establishing formal relations with China before Nixon actually did it.'"

"Bilderberg has had a direct influence on the White House since President Dwight Eisenhower's years," wrote Tucker. "They are accustomed to *owning* the president—whoever he is and whatever party is in control of the White House. A look back: Clinton, Trilateral and Bilderberg; Bush, Trilateral; Regan, received the Trilateralists at the White House and had his Vice President (Bush) address their meeting; Carter [Trilateralist] and his Vice President, Walter Mondale, both Bilderberg; Jerry Ford and his Vice President, Nelson Rockefeller, both Bilderberg..."

Tucker further proclaimed that "Most presidents have been members of at least one of these two groups [Bilderberg or Trilateral Commission], and all have had representatives attend Bilderberg meetings to receive their orders." Allen observed the same pattern; he stated, "While the 'new world order' is being built, the Bilderbergers coordinate the efforts of the European and American power elites." So according to this information the leaders of the NATO nations are covertly controlled by this elite group of financial elitists.

A single currency is said to be in the planning for the entire Western Hemisphere by the Bilderbergers, which will eventually be followed by a world currency. The Bilderbergers, according to Tucker, are also in the process of implementing a global citizens' tax which will be payable to the United Nations. He wrote, "Like the federal income tax, a UN levy would be so small at the outset the consumer would hardly notice. But establishing the principle that the UN can directly tax citizens of the world is important to Bilderberg. It is another giant step toward world government. It is openly discussed with little public notice or objection..."

The objective of the Bilderberg Group appears to be the same as its interlocking Think Tanks, which is a one-world dictatorship, dominated by the financial elite. Journalist Tucker asserts, "Leaders of Bilderberg and its brother group, the Trilateral Commission ... together make up the word shadow government." Tucker informs us that what the Bilderbergers really intend is "a global army at the disposal of the United Nations." And that "it has been a long standing goal of Bilderberg for the UN to become the world government to which all nations will be subservient."

Professor Texe Marrs, has lectured at the University of Texas, authored over 25 books, and has retired from the USAF as an officer with numerous decorations. Writing with authority, backed by careful research, he concludes, "The goal of the Bilderbergers is to create a pretense [fake] democratic World Government, controlled by them."

Author Marrs noted that Bilderberg members that are not part of the inner-core, may not be aware of the true intentions or objectives of the group, and may be convinced that world government is the only way to peace. This also appears to be the situation with its interlocking Think Tanks, the Trilateral Commission and the Council on Foreign Relations.(*) "Not everyone who attends one of the Bilderbergers' secret meetings is an *Insider*," agreed Allen. "Bilderberg policy is not planned by those who attend the conferences, but by the elite steering committee of Insiders."

"Remember, the Bilderbergers are philosophically neither Republican nor Democrat, declared Professor Marrs. "Their only 'party' is filthy lucre. ... Money, in turn, brings them power, and these folks have a never satisfied appetite for raw power." Apparently Gorbachev was influenced by the Bilderbergers and talked about a global union during a speaking tour in the United States. "Gorbachev's speeches could have been dictated by Kissinger after being drafted by Rockefeller ... as content was concerned," charged Tucker. "Gorbachev held out the Bilderberg-Trilateral vision of a UN military force that could invade once-sovereign nations to enforce 'human rights.'" "The New World Order means a new kind of civilization," Gorbachev told the Chicago CFR on May 8, 1991.

Regarding how President Clinton pushed the Bilderberger objective in 1999, Tucker wrote, "It appeared in daily newspapers for all to read, but too many sleep. The developments were celebrated by *The Washington Post's* Bilderberg representative Jimmy Lee Hoagland in a column on June 27, 1999 ... [When he proclaimed], 'The president promises a future in which Americans stand ready to intervene militarily if they can stop wholesale racial or ethnic slaughter "within or beyond other nations' borders" ... He promises a new world order.'" The first American woman to attend a Bilderberg meeting was Hilary Clinton in 1997. Author Marrs senses they may have plans for her in the following years.

Summary

Presidents and prime ministers are *selected* by the Bilderbergers and controlled like puppets. Bilderberg meetings are guarded by the military, SWAT teams, local police and private security firms. Like the CFR, their policy is passed without public approval. Their goal is to create a world government under their control, which will have the illusion of freedom.

Their motives for establishing a global dictatorship using a fake democracy will be expanded upon when we examine a future chapter on psychopathy. Most are unaware of their existence due to a media blackout. Despite attending Bilderberg meetings, mainstream media owners have reported little on this group for over fifty years.

Endnotes

* Authors, Allen, Ross, and Sutton all agree there is an inner-core.

Chapter 6

The Federal Reserve System

Apparently there were constant battles for control of the republic between the international bankers and presidents. Controlling the central bank of a country seems to be a crucial part of installing a dictatorship. Allen wrote, "From the earliest days, the Founding Fathers had been conscious of attempts to control America through money manipulation, and they carried on a running battle with the international bankers."

"Essential to controlling a government is the establishment of a central bank with a monopoly on the country's supply of money and credit," wrote Perloff. Meyer Rothschild is said to have remarked, "Let me issue and control a nation's money, and I care not who writes its laws." Allen concurred, "All those who have sought dictatorial control over modern nations have understood the necessity of a central bank. When the League of Just Men hired a hack revolutionary named Karl Marx to write a blueprint for conquest called The Communist Manifesto, the fifth plank read: "Centralization of credit in the hands of the state... Lenin later said that the establishment of a central bank was ninety percent of communizing a country."

Senator Barry Goldwater described an international banker as one who makes money by extending credit to governments. This is more beneficial to the banker than loaning to an individual because governments borrow much more, and they can guarantee repayment by burdening the people with taxes. Also, instead of debt payment, a banker may receive political influence. Perloff wrote, "No turn of events is more lucrative for an international banker than war—because nothing generates more government borrowing faster."

"Whoever controls the volume of money in our country is absolute master of all industry, and commerce ... and when you realize that the system is very easily controlled, one way or another, by a few powerful men at the top, you will not have to be told how periods of inflation, and depression originate."
-President James Garfield July 2, 1881

One attempt to gain control of the country by way of a central bank was with the Bank of the United States (1816-36), which was abolished by President Andrew Jackson. Jackson warned, "The bold effort the present [Bank of the United States] had made to control the government, the distress it had wantonly produced ... are but premonitions of the fate that awaits the American people should they be deluded into a perpetuation of this institution or the establishment of another like it." Note the *semantic deception* used with the naming of the bank, "Bank of the United States," as if to imply that it was part of the government. This was not the last time this tactic was used.(*)

Senator Goldwater wrote, "In the early years of the Republic ... Jefferson opposed Alexander Hamilton's scheme for the First Bank of the United States, and Andrew Jackson abolished Nicholas Biddle's Second Bank of the United States." "America heeded Jackson's warning for the remainder of the century," wrote Perloff. But the "tide began to turn ... with the linking of European and U.S. banking interest, and the growing in power of America's money barons, such as J.P. Morgan, John D. Rockefeller, and Bernard Baruch."

A German banker named Paul Warburg migrated to the U.S. in 1902. He was an associate of the Rothschilds and became a partner in Kuhn, Loeb, and Company, which was headed by Jacob Schiff. The Schiffs also had ties to the Rothschilds, which went back about a century. Warburg began to lecture widely on the need for a central banking system.

Apparently the international bankers became impatient over the unwillingness of congress to accept a central bank. So using the *Problem-Reaction-Solution* formula, Wall Street deliberately created a panic to force congress to create a central banking system controlled by private interests. This was done to eliminate competition and to seize control of the country by way of the Federal Reserve System.

Perloff wrote, "The Panic of 1907 was artificially triggered to elicit public acceptance of this idea. Snowballing bank runs began after J.P. Morgan spread a rumor about the insolvency of the Trust Company of America." He added, "Tragedy is the mother of new directions. The Panic of 1907 spawned the Federal Reserve."

Allen described J. P. Morgan as "an old hand at creating artificial panics." He stated, "Such affairs were well co-coordinated. Senator Robert Owen, a co-author of the Federal Reserve Act, (who later deeply regretted his role), testified before a Congressional Committee that the bank he owned received from the National Bankers' Association what came to be know as the *Panic Circular of 1893*. It established: 'You will at once retire one-third of your circulation and call in one-half of your loans.'"

On April 25, 1949, *Life Magazine* ran an article entitled, *Morgan The Great*, where historian Frederick Lewis Allen reported that "certain chroniclers have arrived at the ingenious conclusion that the Morgan interests took advantage of the unsettled conditions during the autumn of 1907 to precipitate the panic, guiding it shrewdly as it progressed so that it would kill off rival banks and consolidate the preeminence of the banks within the Morgan orbit."

This deliberately created (problem) caused a predictable panic (reaction), which forced congress to create a commission to investigate other banking options, which resulted in the Federal Reserve (solution). In a *Congressional Record* dated, December 22, 1913, vol. 51, Congressman Charles Lindberg declared, "The Money Trust ... caused the 1907 panic, and thereby forced Congress to create a National Monetary Commission."

Perloff describes the *Money Trust* as, Wall Street monopolists such as Rockefeller, Morgan, Warburg, and Schiff. Heading the National Monetary Commission was Senator Nelson Aldrich, who was under the control of the international bankers. "Aldrich was known as the international bankers' mouthpiece on Capitol Hill," wrote Perloff. There was apparently some bribery that took place too.

The commission spent about two years studying central banks in Europe, and finally the "Federal Reserve became law in December 1913," declared Perloff. The solution was drafted at Morgan's hunting club on Jekyl (sic) Island off the coast of Georgia. "The Fed," stated Allen, "was drafted on Jekyl Island in Georgia by Senator Nelson Aldrich; Henry P. Davison of J. P. Morgan and Company; Frank A. Vanderlip, President of the Rockefeller-owned National City Bank; A. Piatt Andrew, Assistant Secretary of the Treasury; Benjamin Strong of Morgan's Bankers Trust Company; and [Rothschild representative] Paul Warburg." Tucker added "The 'Fed,' … is allied with the Bilderberg group, which is composed of the world's biggest moneychangers—led by the Rockefellers and Rothschilds."

The naming of this bank was another tactic of *semantic deception*; there is no reserve, and it's not federal. The Federal Reserve is privately owned, it makes its own polices and is not subject to the president or congress. Many of its members are from the CFR. "Probably 90% of the US citizens think that the Federal Reserve System is one of the branches of the federal government," Ross said, and added, "most think that it is part of the Treasury Department ... because of the term 'Federal' in its name." "This is no accident," he proclaims, but a "psychological ploy to con the Americans into accepting their deception. It is not 'Federal' and there is no 'Reserve.'"

Perloff added, "Indeed, the Fed is authorized to create money—and thus inflate—at will. According to the constitution, only Congress may issue money or regulate its value. The Federal Reserve Act, however, placed these functions in the hands of private bankers—to their perpetual profit." Senator Goldwater wrote, "The accounts of the Federal Reserve System have never been audited. It operates outside of the control of Congress and through its Board of Governors manipulates the credit of the United States."

Federal Reserve board members serve 14-year terms and are appointed by the president. "Since these positions control the entire economy of the country they are far more important than cabinet positions," said Allen. He continued, "These appointments which should be extensively debated by the Senate are routinely approved." The results he says are "ever-increasing debt requiring ever-increasing interest payments, inflation and periodic scientifically created depressions and recessions."

Also in 1913, the 16th amendment was passed which subjected citizens to a federal tax. "Because income tax has been declared unconstitutional by the Supreme Court in 1895, it had to be instituted by constitutional amendment," wrote Perloff. Again, Senator Nelson Aldrich who was *handled* by the international bankers proposed the amendment. "The man who brought forward the amendment in Congress was the same senator who proposed the plan for the Federal Reserve—Nelson Aldrich," observed Perloff.

Apparently the American people agreed to this new tax because they believed it would have a minimal impact on the middle class and would impede the rich. "Initially, it was nominal," Perloff described, "a mere one percent of income under $20,000—a figure few made in those days." According to Perloff the American people were assured it would never increase.

Another reason is that the public was deceived by clever propaganda that the Federal Reserve would stabilize the economy and prevent future panics. "It did nothing of the kind," declared Perloff. "Not only has our nation suffered through the Great Depression and numerous recessions, but inflation and federal debt—negligible problems before the Fed came into existence—have plagued America ever since."

On June 10, 1932 Congressman Louis McFadden, a former chairman of the House Committee on Banking and Currency warned,

"When the Federal Reserve Act was passed, the people of these United States did not perceive that a world banking system was being set up here. A super-state controlled by international bankers and international industrialists acting together to enslave the world for their own pleasure. Every effort has been made by the Fed to conceal its powers but the truth is—the Fed has *usurped* the government."

Usurp
"1 to seize and hold (as office, place, or powers) in possession by force or without right."
-Merriam-Webster's Dictionary

Congressman Lindberg informed the public in a *Congressional Record* dated December 22, 1913, vol. 51, that "This [Federal Reserve] act establishes the most gigantic trust on earth... When the President signs this act the *invisible government* by the money power, proven to exist by the Money Trust investigation, will be legalized..." He warned, "I have seen these forces exerted during the different stages of this bill" and "The money power overawes the legislative and executive forces of the Nation and of the States."

"If the key to controlling a nation is to run its central bank," asked Perloff, "one can imagine the potential of a global central bank, able to dictate the world's credit and money supply. The roots for such a system were planted when the International Monetary Fund (IMF) and World Bank were formed at the Bretton Woods Conference of 1944. These UN agencies were both CFR creations," he added.

Dr. Johannes Witteveen, former head of the IMF, said in 1975 that the agency should become "the exclusive issuer of official international reserve assets." In the fall 1984 issue of a CFR newsletter called, *Foreign Affairs*, Richard N. Cooper laid out a modern plan for international currency. He wrote, "I suggest a radical alterative scheme for the next century: the creation of a common currency for all the industrial democracies, with a common monetary policy and a joint bank of issue to determine that monetary policy."

In 1987 Senator Jesse Helms, stated "it is no secret that the international bankers profiteer form sovereign state debt. The New York banks have found important profit centers in lending to countries plunged into debt by Socialist regimes. Under Socialist regimes, countries go deeper and deeper into debt because Socialism as an economic system does not work. International bankers are sophisticated enough to understand this phenomenon and they are sophisticated enough to profit from it."

"Many historians would have us believe that this trio of events—the income tax, the Federal Reserve, and the War—was a coincidence," wrote Perloff. "But too often history has been written by authors financed by foundations, in books manufactured by Establishment publishing houses." He observed, "Many more 'coincidences' were yet to trouble the American people in this century."

Summary

Congressmen McFadden and Lindberg, both clearly state that the "legislative and executive forces" had been "usurped" (taken over) by an "invisible government" consisting of international "bankers" and "industrialists" acting together to "enslave the world for their own pleasure." What these researchers and statesmen are saying is that the Federal Reserve is a privately owned corporation, which was created using lies and deceit. It exists to control the nation, and to eliminate competition using taxation, in order to prevent the middle class from rising to affluence.

Endnotes

* A term commonly used by Charlotte Iserbyt in her book, *The Deliberate Dumbing-Down of America.*

Chapter 7

Tax-exempt Foundations

Tax-exempt Foundations were originally setup for humanitarian purposes to provide grants to existing institutions. Rene A. Wormser served as General Counsel to the Reece Committee, which was a congressional committee that investigated the Tax-exempt Foundations from 1953 to 1955. His book, *Foundations: Their Power and Influence*, is a documented expose of his experience with the committee. In it he wrote, "Foundations were originally created to support existing institutions and to undertake certain 'operating' functions."

Soon after (or possibly from their inception) foundations became a loop hole that the financial elite used to avoid taxes. "By the time the income tax became law in 1913, the Rockefeller and Carnegie foundations were already operating. Income tax didn't soak the rich, it soaked the middle class," wrote Perloff. "Because it was a graduated tax, it tended to prevent anyone from rising into affluence. Thus it acted to consolidate the wealth of the entrenched interests, and protect them from new competition."

Smoot pointed out that the primary purpose of some of the large Tax-exempt Foundations is no longer humanitarian in nature, but "predominately tax avoidance." "One of the leading devices by which the wealthy dodge taxes" concurred Perloff "is the channeling of their fortunes into tax-free foundations." He also charged that, "The major foundations, though commonly regarded as charitable institutions, often use their grant-making powers to advance the interests of their founders."

The "independent, uncontrolled financial power often enables foundations to exert a decisive influence on public affairs," wrote Wormser. He further testified that, "They have a power comparable to political patronage." He cautioned "When they do harm, it can be immense harm—there is virtually no counterforce to oppose them."

What other projects do they fund? According to the findings of congressional investigations, the foundations have been known to fund political movements in a direction inclined to favor a socialistic, one-world government. Individual foundations have also been known to merge themselves in a "cartel-like" fashion to fund their political projects, which tends "to endanger the freedom of our intellectual and public life," warned Wormser. Wormser referred to this merging as the *Tax-exempt Complex*.

The first glimpse into foundation influence came under the Congressional Act of August 23, 1912, when the Commission on Industrial Relations studied labor conditions and the treatment of workers by the major U.S. industrial firms. They eventually examined the foundations, which were interlocked with them. "Starting with a study of labor exploitation, it [the Commission on Industrial Relations] went on to investigate concentrations of economic power, interlocking directorates, and the role of the then relatively new large charitable foundations (especially of Carnegie and Rockefeller) as instruments of power concentration," wrote Wormser.

During the commission hearings, future Supreme Court Justice Louis D. Brandeis testified on January 23, 1915, that he was seriously concerned about the emerging danger of such a concentration of power. He said, "When a great financial power has developed ... which can successfully summon forces from all parts of the country ... to carry out what they deem to be their business principle ... [there] develops within the State a *state* so powerful that the ordinary social and industrial forces existing are insufficient to cope with it."

"Control is being extended largely through the creation of enormous privately managed funds for indefinite purposes, hereinafter designated 'foundations'" declared Mr. Basil M. Manly, director of research for the commission. The commission's report concluded that, "As regards the 'foundations' created for unlimited general purposes and endowed with enormous resources, their ultimate possibilities are so grave a menace ... [that] it would be desirable to recommend their abolition."

Congress has declared that these foundations, which can be used to fund *anything*, should be eliminated because they are potentially destructive to the republic. According to Rene Wormser, even though these congressional findings occurred in 1915, the time period is irrelevant—they are still quite important. He stated, "Under totally different economic and social conditions, the findings of 1915 are still significant."

The second investigation into the Tax-exempt Foundations came from the Cox Committee which lasted from 1952 until 1953. Again, fears of subversive political objectives funded by these multi-billion-dollar organizations (acting in concert) were eminent.

On August 1, 1951, a motivated Congressman E. E. Cox (Democrat) of Georgia introduced a resolution in the House of

Representatives to conduct a thorough investigation into the foundations. He asserted, "There are disquieting evidences that at least a few of the foundations have permitted themselves to be infiltrated by men and women who are disloyal to our American way of life. They should be investigated and exposed to the pitiless light of publicity..."

The Cox resolution to investigate the foundations was passed in 1952. Unfortunately, Congressman Cox died during the investigation. The commission met the same fate as the one before it. No actions were taken to prevent the expansion of these foundations, or provide means for future accountability to the public. And as Smoot described it, "the final report of his [Cox] committee (filed January 1, 1953) was a pathetic whitewash of the whole subject."

However, it did still yield some important facts. Part of the final report on January 1, 1953 said some foundations "supported persons, organizations, and projects which, if not subversive in the extreme sense of the word, tend to weaken or discredit [our] system as it exists in the United States and to favor Marxist socialism." Or in other words, they were found to promote Communism.

The third attempt to investigate the foundations lasted from 1953 to 1955, during the Reece Committee hearings. Smoot wrote, "On April 23, 1953, the late Congressman Carroll Reece, (Republican, Tennessee) introduced a resolution proposing a committee to carry on the 'unfinished business' of the defunct Cox Committee. The new committee to investigate tax-exempt foundations ... was approved by Congress on July 27, 1953." Author Perloff added, "For what was probably the ... last time, the CFR came under official scrutiny."

Other organizations which came under investigation included, The American Council of Learned Societies, The National Research Council, the Social Science Research Council, the American Council on Education, the National Education Association, the League for Industrial Democracy, the Progressive Education Association, the American Historical Association, the John Dewey Society, and the Anti-Defamation League.

During the investigation, Norman Dodd, Director of Research for the Reece Committee, was invited to the headquarters of the Ford Foundation by its president, H. Rowan Gaither. Gaither, a member of the CFR, revealed that the Ford Foundation was operating under directives from the White House to use their grant-making power to "make every effort to ... alter life in the United States ... to make possible a comfortable merger with the Soviet Union." Apparently Mr.

Dodd was put under surveillance, stalked, and experienced character assassination.

The committee was attacked viciously and resulted in a whitewash. Recognizing another unsuccessful attempt to scrutinize the interlocks, Smoot said, "It went out of existence on January 3, 1955, having proven, mainly, that the mammoth tax-exempt foundations have such power in the White House, in Congress, and in the press that they are quite beyond the reach of a mere committee of the Congress of the United States."

But the committee did yield some helpful information. It found that the Tax-exempt Foundations, their intermediaries and interlocks have "exercised a strong effect" on "public education," which "has been accomplished by [using] vast propaganda, by supplying executives and advisors to government and by controlling much research in this area through the power of the purse." And that, "The net result of these combined efforts has been to promote a 'world government.'"

Quoting from the final report of the committee, Perloff wrote, "The report ... observed that major foundations have actively supported attacks upon our social and government system and financed the promotion of socialism and collectivist ideas." The Committee declared that the CFR was "in essence an agency of the United States Government" and that its "productions are not objective but are directed overwhelmingly at promoting the globalist concept."

"The Reece Committee ... proved with an overwhelming amount of evidence that the various Rockefeller and Carnegie foundations have been promoting socialism since their inception," agreed Allen. The Reece Committee hearings also revealed that individual Tax-exempt Foundations often act in concert with each other in order to amplify the enactment of their goals.

Congressman Reece made a final report on Tax-exempt Foundations, which was published by the government printing office on December 16, 1954. He said that there was clearly an interlock between The Carnegie Endowment for International Peace, and some of its associate organizations, such as the Council on Foreign Relations and other foundations, with the State Department. And that, "[the] foundations and organizations would not dream of denying this interlock. They proudly note it in reports."

Reece details the infiltration into the government, stating, "They [CFR/foundation interlock] have undertaken vital research projects for the [State] Department ... [and have] fed a constant stream of personnel into the State Department trained by themselves or under programs which they have financed."

Finally Reece concluded that "the Rockefeller Foundation, The Carnegie Corporation of New York, and the Carnegie Endowment for International Peace, [are] using their enormous public funds to finance a one-sided approach to foreign policy and to promote it actively ... by propaganda, and in the Government through infiltration. The power to do this comes out of the power of the vast funds employed."

The report clearly states that the CFR with interlocked foundations have infiltrated the government. And that they use enormous sums of money to propagandize "educate" the public in support of the policies which they decide we should adopt. What congress has told us is that the government has been infiltrated by the CFR using multi-billion-dollar private bank accounts known as Tax-exempt Foundations.

This means that when a major U.S. policy is filtered down from the federal government, into the local and state governments, that it may originate from the CFR/foundation interlock. Specifically, this means that it has come from the big corporations and the international banks, of which the CFR and other Think Tanks are composed.

The "State within a state" that Justice Louis D. Brandeis warned about in 1915, was part of the beginning of a government within a government, or, as FBI Agent Dan Smoot calls it, *The Invisible Government*. Other notable components include: the emergence of the Federal Reserve System, and the creation and infiltration of the CFR, the Bilderbergers and the TC into the executive branch.

The Reece Committee found that, "When their activities spread into the field of the so-called, 'social sciences' or into other areas which our basic moral, social, economic, and governmental principles can be vitally affected, the public should be alerted to these activities and be made aware of the impact of foundation influence on our accepted way of life."

"The power of the individual large foundations is enormous," they concluded. "It can exercise various forms of patronage which carry with them elements of thought control. ... It is capable of invisible coercion through the power of its purse. ... This power to influence national policy is amplified tremendously when foundations act in concert. There is such a concentration of foundation power in the united States."

"Every significant movement to destroy the American way of life has been directed and financed, in whole or in part, by tax-exempt organizations, which are entrenched in public opinion as benefactors of our society," warned Smoot. And Wormser stated, "By engaging 'public relations counselors' (ethically, and even legally, a questionable practice), it can further create for itself a favorable press and enthusiastic publicity."

"As I see it, the foundations ... have, nonetheless, become the 'agencies' of the principal organization which they finance—the Council on Foreign Relations," said Smoot. So according to Smoot's conclusion, which is backed up by two congressional investigations, the tax-exempt giants are basically the private bank accounts of the interlocking Think Tanks.

Perloff arrived at the same conclusion when he noted, "The Rockefeller Foundation, for example, has poured millions into the Council on Foreign Relations, which in turn serves as the Establishment's main bridge of influence to the U.S. government." Wormser wrote that "Dr. Hutchins ... [former] President of The Ford Foundation's off-shoot, The Fund for the Republic, stated in 1948 ... that 'world government is necessary, therefore it is—or must be made—possible.'" Finally, Smoot summarized, "[foundations] do finance the vast, complex, and powerful interlock of organizations devoted to a socialist one-world system."

Writing about the results of other independent organizations which have investigated foundations, Wormser observed that, "ideas and organizations [supported] by tax-exempt foundations ... had become the breeding ground for socialist and related political movements and action." And that there were fears, "over the danger of foundation support of various undesirable concepts and movements having political implications." These fears included, "the impairment of our national sovereignty; and even subversion. Hence the support by a majority in Congress of both the Cox and Reece Committee inquires."

The Reece Committee also found that foundations tend to support "moral relativity" (the end justifies the means), and "social engineering" (mind-control), which are "detriment of our basic moral, religious, and governmental principles." Wormser indicated that they also supply grants to intermediary organizations, which they've created, in order to carry out private (political) projects. These organizations are essentially *public front organizations*, which are presented to the public as being humanitarian in nature.

Interestingly, the brutal MKULTRA experiments, which were carried out in prestigious hospitals and universities, were funded by the

Rockefeller Foundation. In his book, *The Search for the Manchurian Candidate*, Jonathan Marks wrote, "He [Dr. Cameron] headed Allan Memorial since 1943, when the Rockefeller foundation had donated funds to set up a psychiatric facility at McGill University. With continuing help from the Rockefellers, McGill had built a hospital known far beyond Canada's borders..."

Summary

So, as these investigations and independent researchers have found, the true purpose for some of these foundations which are used by the financial elite, is to bankroll the installation of a one-world socialist dictatorship, and tax avoidance. These investigations were launched because Congress and others were concerned that these foundations were backing subversive socialist political moments by using vast propaganda and the power of the purse.

Chapter 8

Commentary

There can be no denying that these organizations have infiltrated the executive branch. According to this information, it began with the documented overthrow of the true government in 1913 and continued with the appointment of members of elite Think Tanks into the executive branch.(*) First with the CFR, then the Bilderbergers and finally the Trilaterals. The infiltration by the Tax-exempt Foundations and the interlocking Think Tanks was done by the "power of the purse," according to the Reece Committee hearings.

There is a heavy cross membership between those in control of the foundations and the elite societies. The Tax-exempt Complex is essentially their private multi-billion-dollar bank account. There is a strong interlock between the elite societies, the owners of mainstream media, the foundations, the Federal Reserve, Wall Street, the multinational corporations, and international banks. This cabal controls the United States and other NATO governments.

Once the deciding members of these Think Tanks settle on a policy, significant resources are pooled to promote the policy, which filters down to the local government. Opposition is met with ridicule by an array of organizations which the elite use to confound and discredit their critics. And in their own publications, they've made their intentions perfectly clear—a removal of the constitution and individual freedom, and the creation of a one-world government with them in control. This will be an absolute *nightmare* of a dictatorship.

On August 30, 2003, Congressman Ron Paul was addressing an audience near Austin Texas and was asked if there existed a "deceptive conspiracy to overthrow the American Republic" and "Bill Of Rights" and to "usher in a totalitarian World Government," under the control of the UN. He answered, "Yes. I think there are 25,000 individuals that have used offices of power, and they are in our Universities and they are in our Congress, and they believe in One World Government. And if you believe in One World Government, then you are talking about undermining National Sovereignty and you are talking about setting up something that you could well call a Dictatorship."

Obviously, no presidential candidate is allowed to get close to the White House without being *owned* by this cabal. Although voting fraud probably does exist, the elections themselves are apparently scams, which offer the illusion of choice. This explains why there have been no major positive changes throughout multiple administrations.

"Single acts of tyranny may be ascribed to the accidental opinion of the day; but a series of oppressions, begun at a distinguished period, and pursued unalterably through every change of... [administrations] too plainly proves a deliberate, systematic plan of reducing us to slavery." -Thomas Jefferson

In his book, *How the World Really Works*, Alan B. Jones wrote, "the elites who wish to set up their world dictatorship recognize their main enemy to be the great middle class of the United States, which, being made up of individuals who have acquired a little education, property, and independence, will fight strenuously to keep them."

"The strategy," says Jones, "of the elites is to squeeze the middle class to death by creating or exacerbating the major problems facing the society, including class warfare, crime, education, moral decay, etc., and then creating in response spurious governmental programs to "cure" the problems that they just created." What Jones is describing here is the financial elite's consistent use of the *Problem-Reaction-Solution* formula.

"I have two great enemies, the Southern Army in front of me and the Bankers in the rear. Of the two, the one at my rear is my greatest foe." -Abraham Lincoln

Now that you have an understanding of the real power behind the throne, in upcoming chapters we'll explore how these organizations have manipulated some major world events. But first let's examine their control of mainstream information more closely. This will help explain why you may never have heard of them, or the authentic version of how some historical events occurred.

Endnotes

* Former presidents, such as Garfield and Lincoln have attested to having running battles for the republic with the international banks and corporations. So some researchers may contend that this overthrow occurred at an earlier date, and they may be correct. Other authors suggest that despite winning the Revolutionary War, the United States was never truly removed from the grip of British Royalty. They too may be correct. But I listed 1913 as it was documented as an overthrow in a *Congressional Record* dated December 22, 1913, vol. 51, and echoed by Congressman Louis McFadden. This evidence was provided in the Federal Reserve section.

Volume I Part II

Chapter 9

Centralized Control of History, Media, & Academia

Control of Media

In his book, *The Rockefeller File*, free-lance journalist and bestselling author Gary Allen declared, "Rockefeller gang's plans for monopolistic World Government are never, but never, discussed in the machines of mass misinformation." By now you probably already know why you haven't heard of the Hidden Evil, or the coming socialist dictatorship in mainstream news. The people who are installing it control the media. "Why don't investigative news shows like Sixty Minutes ... tackle the ... Trilateralist-CFR hold on our government," asks Perloff. "Surely such material would have sufficient audience appeal."

"The so-called founders of such giants as *The New York Times* and *NBC* were chosen, financed and directed by Morgan, Schiff, and their allies," wrote Allen. Regarding the beginning of the takeover of the mainstream news, Perloff says, "A prime mover in this process was J.P. Morgan—the original force behind the CFR." Perloff quotes Congressman Callaway's notation in *Congressional Record* volume 54, dated February 9, 1917, which stated, "In March, 1915, the J.P. Morgan interests ... got together 12 men high up in the newspaper world and employed them to select the most influential newspapers ... to control generally the policy of the daily press of the United States."

The record continued, "These 12 men ... found it was only necessary to purchase the control of 25 of the greatest papers. ... Emissaries were sent to purchase the policy, national and international, of these papers ... [and] an editor was furnished for each paper to properly supervise and edit information regarding ... things of national and international nature considered vital to the interest of the purchasers. ... This policy also included the suppression of everything in opposition to the wishes of the interests served."

"The CFR needs to reach the mass audience of Americans who do not belong to, or attend the meetings of, or read material distributed by, the propaganda organizations," declared Smoot. He continued, "[The] Council on Foreign Relations leaders are aware of this need, and they have met it." Allen agreed, when he wrote, "Public opinion is manufactured by the CFR's ventriloquists in the mass media."

Former CFR member, Admiral Chester Ward, described the CFR's influence in the mass media, too. He proclaimed, "They control or own major newspapers, magazines, radio and television networks," and "the most powerful companies in the book publishing business." *The Associated Press*, *New York Times*, *Washington Post*, and *LA Times* have their own wire services, which most mainstream news outlets use. According to Allen, the CFR controls these wire services.

Ross observed, "The CFR could not accomplish their goals without complicity of the mainstream news media which they absolutely control with an iron fist." [Emphasis in original] He explained that, "Occasionally they will hold a public meeting, and invite the open press (including *C-SPAN*), in order to give the impression that they are a harmless group engaged only in social activities."

Allen refers to the mainstream media as "CFR's ventriloquists," and writes, "At the center of Insider power, influence, and planning in the United States is the pervasive Council on Foreign Relations." He adds, "The CFR was created by the Rockefellers and their allies to be the focus of their drive for a 'New World Order.'"

"We are grateful to The Washington Post, the New York Times, Time Magazine and other great publications whose directors have attended our meetings and respected their promises of discretion for almost forty years. It would have been impossible for us to develop our plan for the world if we had been subjected to the lights of publicity during those years. But, the world is now more sophisticated and prepared to march towards a world government. The supranational sovereignty of an intellectual elite and world bankers is surely preferable to the national auto-determination practiced in past centuries."
-David Rockefeller, Bilderberg Meeting, Baden-Baden Germany, 1991

Other outlets allegedly under the control of the CFR include, the *National Broadcasting Corporation*, *Columbia Broadcasting System*, *Time*, *Fortune*, *Look*, *Newsweek*, *New York Post*, *Denver Post*, *Louisville Courier Journal*, *Minneapolis Tribune*, and the *Knight Papers*. Publishing houses under CFR control include, *McGraw-Hill*, *Simon & Schuster*, *Harper Bros.*, *Random House*, *Little Brown & Co.*, *Macmillan Co.*, *Viking Press*, *Saturday Review*, *Business Week*, and *Book of the Month Club*.

The mainstream media appears to serve multiple functions. First, to distribute information that the controlling cabal wants you to believe is real news. Secondly, to assist in the cover-up of information that has leaked out through other sources. In other words, *Damage Control.* Thirdly, as a weapon to assassinate the characters of those who expose the practices of the Establishment. Also, to act as a firewall to prevent information detrimental to the elite's control from reaching the public. Finally, to continue to reinforce mainstream accounts of current and historical events. The mainstream news essentially serves as a primary tool for social conditioning or *mind-control.*

"It is through the press and the media that the lie is penetrating through to the masses," declared Professor Marrs. The media is used as a weapon. So when you watch, read, or listen to mainstream news, ask yourself, "what will the CFR like me to believe today?" Allen sums up the real purpose of the mainstream news, stating, "These are the Establishment's official landscape painters whose jobs it is to make sure the public does not discover the C.F.R. and its role in creating a world socialist dictatorship."

According to Allen, "psychology and propaganda" used in the mainstream news, were "developed in the West in such places as the Rockefeller financed Tavistock Institute in England." He says, "The hidden persuaders from Madison Avenue, the Rand Corp. think tank or Hudson Institute, can and do manipulate public opinion." "The Establishment elitists," he explains "refer to it as 'the engineering of consent.' That means we are made to think the manacles they are slipping on our wrists are love bracelets." Allen charges that the techniques developed at these Think Tanks have been used successfully by Communist states.

"Think of the press as a great keyboard on which the government can play."
-Dr. Joseph Goebbels, Nazi Propaganda Minister

From the descriptions that these researchers have given, an accurate analogy could be drawn from the Wizard of Oz. America and other NATO are *the great and powerful Oz*, which appear to be run by their respective leaders. The mainstream news is the "curtain," which prevents us from understanding that the real controllers behind it are Think Tanks, the Fed, Wall Street, and the Tax-exempt Complex.

Another role of the mainstream news is to project the illusion of a two party system. The grass roots Democrats and Republicans have their philosophical differences. But, says Allen, "as you move up the party ladders these differences become less and less distinguishable until finally the ladders disappear behind the Establishment's managed news curtain and come together at the apex under the control of the C.F.R."

"There is no such thing at this date of the world's history, in America, as an independent press. You know it and I know it... The business of journalists is to destroy the truth, to lie outright... We are the tools of rich men behind the scenes. We are jumping jacks, they pull the strings and we dance. Our talents, our possibilities and our lives are all property of other men. We are intellectual prostitutes."
-Journalist John Swinton, New York Press Club 1953

Most people pay attention to the events illuminated by these corporate-owned spotlights because they're big, common, socially accepted, and because they're all most people have ever known. When you start listening to independent news broadcasts, your conditioning begins to come undone, you start to connect the dots and find that events that were puzzling and seemed unrelated are part of the same picture. These independent news outlets broadcast on shortwave, and web cast. Some have past MP3 shows available for download. These shows are extremely educational.(*)

By listening to these broadcasts you can become familiar with who really runs this planet, what they've done, and where they're taking you. Most of the topics discussed on these shows will never be covered on mainstream news. If you do begin to listen to these shows, you may find yourself watching mainstream news less frequently due to its high rubbish content. You may also become temporarily upset after realizing you've been deceived, you've been misled, conned, tricked—and that most of your friends and relatives have too.

Control of History & Education

"According to some, history is basically a jumble of events: blunders, coincidences, and happenstances that have brought us to where we are today," wrote Perloff. "However, seen in the context of globalist influence and maneuvering, history ... begins to make sense, as if snapping into place upon a calculated blueprint."

In 1915, Mr. Basil M. Manly, director of research for a congressional investigation known as the Commission on Industrial Relations, observed that control of information was being extended by, the "endowment of colleges and universities, [and] through controlling or influencing the public press," wrote Wormser. During the Reece Committee hearings, Professor Kenneth Colegrove at Northwestern University testified, "The officers of these foundations wield a staggering sum of influence and direction upon research, education and propaganda in the United States and even in foreign countries."

The Reece Committee concluded, "The impact of foundation money upon education has been very heavy, largely tending to promote uniformity in approach and method, tending to induce the educator to become an agent for social change and a propagandist for the development of our society in the direction of some form of collectivism. Foundations have supported text books ... which are destructive of our basic governmental and social principles and highly critical of some of our cherished institutions."

Wormser exposed the profound influence that foundations have on public education, when he stated, "Foundation giving most obviously has an enormous impact on education, on social thinking, and ultimately on political action. This influence reaches the public through the schools and academies, through publicity, and through education and other associations dedicated to public and international affairs." He sustained, "Foundation grants have become so important a source of support that college and university presidents cannot often afford to ignore the opinions and wishes of the executives who distribute foundation largess."

Finally, he added, "This situation permits large foundations to exercise a profound influence upon public opinion." Senator Goldwater agreed, writing, "The intellectual influence of the academic community ... is subservient to the wealth of the great tax-free foundations."

Observing the effort by the Establishment to lie about history, former FBI Agent Dan Smoot said, "The Reece Committee investigation threw some revealing light on the historical blackout which the Council on Foreign Relations has ordered and conducted." Smoot gave one example which included a two-volume history of World War II, written in part by William L. Langer of the CFR and funded by grants from the Rockefeller Foundation and the Alfred P. Sloan Foundation.

"Foundations have supported a conscious distortion of history," revealed the Reece Committee. The phrase "conscious distortion" is a polite way of saying *blatant lie*. "Twentieth-century history, as recorded in Establishment textbooks and journals, is inaccurate," stated Professor Sutton. "The prevailing Establishment version is seen to be, not only inaccurate, but designed to hide a pervasive fabric of deceit and immoral conduct." Professor Sutton explained that they are effective, "Not so much because of outright censorship, although that is an important element, but more because of the gullibility of the American 'educated public.'"

Professor Marrs concludes that these historical lies are deliberately manufactured so the public will be easier to control. He charges, "The rewriting and deceitful misinterpretation of American and world history is one of the main methods used to manipulate and shape public opinion." "The establishment of the American Historical Society was no doubt an unprecedented step in instituting this diabolical plan to rewrite the past," he observes.

The accrediting agencies themselves may also be controlled by the Tax-exempt Foundations, as suggested by the Reece Committee. It stated that the "so-called "accrediting' organizations ... are extra-governmental, yet ... [influence] education to a considerable degree. For various reasons colleges, universities, and specialized schools and departments today require 'accreditation,' that is, approval of one or more of these organizations which presume to set standards. Some of these accrediting organizations are supported by foundations..."

"Universities, hospitals, institutes and learned societies sometimes supply nothing but their name labels affixed to what is actually a pet project of foundation managers," announced Wormser. According to Wormser, they pick the professors, they pick the topics, and they even pick the research staff. And they already know what the outcome is going to be. These doctors, professors, research staff, etc. are made well aware in advance that there will be a particular conclusion that they must arrive at. Not adhering to this predetermined conclusion will result in a denial of grants.

Professor Sutton agreed, writing, "Through foundations controlled by this elite, research by compliant and spineless academics, 'conservatives' as well as 'liberals' has been directed into channels useful for the objectives of the elite essentially to maintain this subversive and unconstitutional power apparatus. Through publishing houses controlled by this same financial elite unwelcome books have been squashed and useful books promoted."

"It is understandable," continued the professor, "that universities and research organizations, dependent on financial aid from foundations that are controlled by this same New York financial elite, would hardly want to support and to publish research on these aspects of [authentic] international politics." Books critical of the official account of history don't get reviewed or are not available through mainstream outlets. And according to author Perloff, historians that do not adhere to their version of history are often *blacklisted.*

Smoot asks, "Where will you find a college administration that will not defend the Ford Foundation against all critics—if the college has just received, or is in line to receive, a million-dollar gift from the Foundation? How far must you search to find college professors or school teachers who will not defend the Foundation which gives 25 million dollars at one time, to raise the salaries of professors or school teachers?"

When I present evidence of some authentic history in the next chapter, you'll also notice a pattern of attacks against people who tell the truth. These attacks are launched using mainstream media and government "experts." "During the past one hundred years any theory of history or historical evidence that falls outside a pattern established by the American Historical Association and the major foundations with their grant-making power has been attacked or rejected," wrote Professor Sutton.

In a future chapter called, *The Satanic Influence*, we'll find evidence that the AHA was setup by an Establishment organization known as *The Order of Skull and Bones.* This organization, like its elite Think Tank counterparts, works toward the objective of a one-world socialist dictatorship under elite control. The mainstream media is used as a weapon by these groups.

Allen, observed that the Rockefellers "do not pour money into local school board races; they put their bucks into the schools that train the teachers and they finance the writing of textbooks." He stated, "The family couldn't care less who controls the local school board." According to Allen, the publishing houses under CFR control also specialize in publishing school textbooks.

Speaking of the movement by the Council on Foreign Relations to use professors in their deliberate campaign to rewrite the past, author Perloff noted, "The council's [CFR's] steering committee moved to distinguish the roster by adding college professors. ... They hailed from campuses beholden to J.P. Morgan. As Dr. Quigley observed: 'The Wall Street contacts with these professors were created originally from Morgan's influence in handling large academic endowments.'"

This evidence suggests that mainstream schools and media are conduits for historical lies. It also suggests that they are a medium for social conditioning, also known as *mind-control.* Professor Sutton observed, "The existing system of education is little more than a conditioning mechanism. It has little to do with education in the true sense, and a lot to do with control of the individual. This requires suppression of individualist tenancies and a careful spoon-feeding of approved knowledge."

Professor Marrs declared, "Few of us are aware that such techniques are being utilized against us since we're constantly being led to believe ... that we are independent, resourceful, intelligent, and fully able to make up our own minds." Many of us have operated most of our lives on a filtered version of history. Or, in essence, lies.(**)

Summary

So, according to the results of congressional investigations and independent researchers, mainstream history is written by the financial elite. The CFR and interlocking Think Tanks have literally rewritten history using schools and publishing houses which they control by grants from their private multi-billion-dollar bank accounts (Tax-exempt Foundations).

I'm not an expert on history, but the information I'll be presenting in future chapters will be more accurate than some mainstream accounts. Now that you understand that a financial elite does exist and that they exercise an enormous amount of control over mainstream publications, it will be easier to understand why the following events did not unfold as presented in schools and mainstream media. Although some of this information pertains directly to the U.S., it is relevant to all NATO countries, because as goes America, so goes the rest of the world.

Endnotes

* I would highly recommend the *Republic Broadcasting Network* (RBN), at www.republicbroadcasting.org, and the *Genesis Communications Network* (GCN), at www.gcnlive.com.

** I include myself in this category, researching information for this book has been a learning and sometimes shocking experience. And there is still much to learn.

Chapter 10

Wall Street Funded Communists

Professor Sutton stated, "Western textbooks on Soviet economic development omit any description of the economic and financial aid given to the 1917 Revolution and subsequent economic development by Western Firms and banks." "In the Bolshevik Revolution we have some of the world's richest and most powerful men financing a movement which claims its very existence is based on the concept of stripping of their wealth," declared Allen. "[M]en like the Rothschilds, Rockefellers, Schiffs, Warburgs, Morgans, Harrimans, and Milners."

Perloff agreed, "Jacob Schiff, the head of Kuhn, Loeb and Co., heavily bankrolled the [Communist] revolution. This was reported by White Russian General Arsine de Goulevitch in his book *Czarism and the Revolution*." "According to his grandson John," described Allen, "Jacob Schiff ... long-time associate of the Rothschilds, financed the Communist Revolution in Russia to the tune of $20 million." He continued, "According to a report on file with the State Department, his firm, Kuhn Loeb and Co. bankrolled the first five year plan for Stalin," and added, "Schiff's descendents are active in the Council on Foreign Relations today."

Referring to the emergence of a communist dictatorship which resulted from the Bolshevik Revolution in 1917, Professor Marrs wrote that they were funded by "Germany and America. ... Their repugnant campaign to purify and cleanse Mother Russia and to seek world domination resulted in ... [millions of] human beings wiped out and brutally purged..." He attested, "Brown Brothers Harriman" helped finance it with "money made possible by it and the affiliated Guaranty Trust Company." Professor Sutton agreed, writing "W. Averell Harriman was a director of Guaranty Trust Company" and "was involved in the Bolshevik Revolution."

On February 3, 1949, the *New York Journal-American* stated, "Today it is estimated even by Jacob's grandson, John Schiff, a prominent member of New York Society, that the old man sank about $20,000,000 for the final triumph of Bolshevism in Russia. Other New York banking firms also contributed."

Referring to a June 15, 1933, *Congressional Record*, Allen wrote "Congressman Louis McFadden, chairman of the House Banking Committee, maintained in a speech to his fellow Congressman: "The Soviet government has been given United States Treasure funds by the

Federal Reserve Board and the Federal Reserve Banks acting through the Chase Bank and the Guaranty Trust Company and other banks in New York City. ... Open up the books of Amtorg, the trading organization of the Soviet government in New York, and of Gostorg, the general office of the Soviet Trade Organization, and of the State Bank of the Union of Soviet Socialist Republics and you will be staggered to see how much American money has been taken from the United States' Treasury for the benefit of Russia."

"Now our textbooks tell us that the Nazis and Soviets were bitter enemies and their systems are opposites," observed Professor Sutton. But in the "1920s, W. Averell Harriman was a prime supporter of the Soviets with finance and diplomatic assistance ... [and] participated in RUSKOMBANK," which was "the first Soviet commercial bank. Furthermore, Max May, the Vice President of Guaranty Trust, which was dominated by Harriman and Morgan, became the first Vice-President of Ruskombank." However, declared Professor Sutton, "Averell Harriman, his brother Roland Harriman, and ... E.S. James and Knight Woolley, through the Union Bank ... were prime financial backers of Hitler." He asked, "How could a rational man support Soviets and Nazis at the same time?"

A curious dilemma arises when faced with the documented fact that Wall Street funded both Communists and Nazis. First, it would seem these two forms of government are at opposite ends of the political spectrum. And that capitalists would see them as a threat to their growth. Allen provides a possible answer, writing, "But obviously these men have no fear of international Communism. It is only logical to assume that if they financed it, and are willing—even eager—to cooperate with it, it must be because they control it. Can there be another explanation that makes sense?" He adds, "Remember that for over 100 years it has been a standard operating procedure of the Rockefellers and their allies to control both sides of every conflict."

Before Winston Churchill became Prime Minister of Great Britain, he acknowledged a conscious effort of wealthy people to install a communist dictatorship in Russia. He wrote in the February 18th, 1920 issue of the London Illustrated Sunday Herald, that "From the days of Spartacus [Adam] Weishaupt ... to those of Karl Marx, to those of Trotsky, Bella Kuhn, Rose Luxembourg, and Emma Goldman, this world-wide conspiracy has been steadily growing." He affirmed, "It has been the mainspring of every subversive movement during the 19th century; and now at last, this band of extraordinary personalities from the underworld of the great cities of Europe and America have gripped the Russian people by the hair of their heads and have become practically undisputed masters of that enormous empire."

After the Bolshevik Revolution, Wall Street ensured the communists would retain control of Russia. Professor Sutton described this effort when he wrote, "On may 1st, 1918, when the Bolsheviks controlled only a small fraction of Russia (and were to come near to losing even that fraction in the summer of 1918), the American League to Aid and Cooperate with Russia was organized in Washington, D.C to support the Bolsheviks. This was not a 'Hands off Russia' type of committee formed by the Communist Party U.S.A or its allies. It was a committee created by Wall Street with George P. Whalen of Vacuum Oil Company as Treasurer and Coffin and Oudin of General Electric, along with Thompson of the Federal Reserve System, Willard of the Baltimore and Ohio Railroad, and assorted socialists."

"The Bolsheviks were not a visible political force at the time the Czar abdicated," Allen wrote. "And they came to power not because of the downtrodden masses of Russia called them back, but because very powerful men in Europe and the United States sent them in. "They [Lenin and Trotsky] joined up, and, by November, though bribery, cunning, brutality and deception, they were able (not to bring the masses rallying to their cause, but) to hire enough thugs and make enough deals to impose *out of the gun barrel* what Lenin called 'all power to the Soviets.'"

"Having created their colony in Russia," said Allen, "the Rockefellers and their allies have struggled mightily ever since to keep it alive. Beginning in 1918 this clique has been engaged in transferring money and, probably more important, technical information to the Soviet Union." Perloff agreed, writing, "Probably no name symbolized capitalism more than Rockefeller. Yet that family has for decades supplied trade and credit to Communist nations. After the Bolsheviks took power, the Rockefellers' Standard Oil of New Jersey bought up Russian oil fields, while Standard Oil of New York built the soviets a refiner and made an arrangement to market their oil in Europe. During the 1920's the Rockefellers' Chase Bank helped found the American-Russian Chamber of Commerce, and was involved in financing Soviet raw material exports and selling Soviet bonds in the U.S."

According to Senator Barry Goldwater, Chase Manhattan built a truck factory in Russia which could also be used to produce armored vehicles such as tanks and even rocket launchers. Perloff echoed, "American technology helped the Soviets construct the $5 billion Kama

River truck factory ... [which was] successfully converted by the Kremlin to military purposes."

Wall Street continued to aid the Russian communists as they supplied the Vietnamese communists that Americans were fighting in Vietnam, says Allen. In the late 60s Rockefeller and other industrialists built synthetic rubber plants and an aluminum factory totaling about 250 million dollars. Professor Sutton observed, "these American capitalists were willing to finance and subsidize the Soviet Union while the Vietnam War was underway, knowing that the Soviets were supplying the other side."

An article appeared in the New York Times on January 16, 1967, which carried the headline, *Eaton Joins Rockefellers to Spur Trade with Reds*. Perloff summed up the story, "The ensuing story noted that the Rockefellers were teaming up with tycoon Cyrus Eaton, Jr., who was financing for the Soviet block the construction of a $50 million aluminum plant and rubber plants valued at over $200 million." He added, "The Chase, which maintains a branch office at 1 Karl Marx Square in Moscow, has gained notoriety for financing projects behind the Iron Curtain."

W. Averill Harriman was made U.S. Ambassador to the USSR in 1941. Allen wrote, "Sutton quotes a report by Averell Harriman to the State Department in June, 1944 as stating: 'Stalin paid tribute to the assistance rendered by the United States to Soviet industry before and during the war.'" "It is not an exaggeration to say that the USSR was made in the USA," observed Allen. Referring to Professor Sutton's book, *Wall Street and the Bolshevik Revolution*, he wrote, "No one has even attempted to refute Sutton's almost excessively scholarly works. They can't. But the misinformation machines that compose our mediacracy can ignore Sutton. And they do." The book, added Perloff, "is based on assiduous research, including a deeper probe into State Department files."

Professor Sutton warned, "The synthesis sought by the Establishment is called the New World Order. Without controlled conflict this New World Order will not come about. ... And this is being done with the calculated, managed, use of conflict. ... This explains why the International bankers backed the Nazis, the Soviet Union, [and] North Korea ... against the United States. The 'conflict' ... [builds] profits while pushing the world ever closer to One World Government. The process continues today."

Summary

The evidence that these researchers present, which includes mainstream news and State Department records, suggests that Communist Russia was not only heavily funded by Wall Street, but was the actual creation of Wall Street. The Communist Revolution was instigated by Wall Street. Wall Street continued to build Russia even as they supplied aid to a country that America was at war with. Because they created communist Russia, they must have known about the millions of people being murdered.

Chapter 11

Pearl Harbor was Allowed to Happen

The following evidence suggests that the attack on Pearl Harbor was allowed to happen to justify the entry of America into WWII. At this point in history, the CFR had infiltrated the White House and was in control of FDR. "The Council on Foreign Relations played a significant role in the Roosevelt administration," declared author Perloff. This was done through the War and Peace Studies Project, which was funded by the Rockefeller Foundation. For instance, Henry Stimson was War Secretary for FDR, and George Messersmith was Secretary of State. So when you see FDR, think "CFR."

U.S. intelligence cracked the radio code that Tokyo used to communicate with its embassies. The decoded messages, which were usually known to Washington on a daily basis, revealed that spies in Hawaii informed Tokyo of the exact locations of warships at Pearl Harbor. Washington was also aware that the attack would come on or around December 7th.

Perloff observed, "These intercepts were routinely sent to the President and to Army Chief of Staff General George Marshall. In addition, separate warnings about the attack—with varying specificity as to its time—were transmitted to these two men by or through various officials, including Joseph Grew, our ambassador to Japan; FBI Director J. Edgar Hoover; Senator Guy Gillette ... Congressman Marin Dies; Brigadier General Elliot Thorpe ... Colonel F.G.L. Weijerman ... and other sources."

On December 6th, Captain Johan Ranneft of the Dutch naval attaché in Washington acknowledged that U.S. naval intelligence informed him that Japanese carriers were about 400 miles northwest of Honolulu. Apparently Ranneft informed Washington, but no alert was passed to the commanders in Hawaii. At least on hour before the attack, a Japanese submarine was sunk at the mouth of the harbor by a US Navy destroyer called the *USS Ward*. The captain of the *USS Ward*, Lieutenant William W. Outerbridge, sent a message to Admiral Kimmel. But it seems that there was some communications trouble between Outerbridge and Kimmel.

The Times (UK), wrote, "Outerbridge sent a report of the attack to Admiral Husband E. Kimmel, commander in chief of the Pacific Fleet." Outerbridge's report to the US Navy stated that, "Damage was seen by several members of the crew. This was a square positive hit ...

[and the] submarine was seen to keel over to starboard." The sighting of the submarine was confirmed by a minesweeper called the *USS Condor*. But Admiral Kimmel decided to "wait for verification of the report" before raising the alarm to the rest of the US Pacific Fleet in Pearl Harbor, they reported.

Prior to the attack, Admiral Richardson was stripped of his command of Pearl Harbor by FDR, for warning of the fleet's vulnerability. Command of the base was given to Admiral Kimmel. Perloff commented, "Kimmel's predecessor, Admiral Richardson, had been removed by FDR after protesting the President's order to base the Pacific Fleet in Pearl Harbor, where it was quite vulnerable to attack." He continued, "Roosevelt and Marshall [also] stripped the island of most of its air defense shortly before the raid, and allotted it only one third of the surveillance planes need to reliably detect approaching forces."

An article in the *BBC* dated August 29, 2002, entitled *Japanese Pearl Harbor Sub Found*, stated, "The 78-foot (24-metre) submarine could provide the first physical evidence to back US claims that it fired first against Japan in World War II and inflicted the first casualties." Also on August 29, 2002, the *Chicago Sun-Times* ran an article entitled *Japanese Sub Sunk Before Attack Found in Pearl Harbor*. It declared, "Researchers said Wednesday they found a Japanese midget submarine sunk nearly two hours before the aerial attack on Pearl Harbor." And *The Times* reported, "Daniel Martinez, a historian for the *USS Arizona Memorial* who has interviewed the sailors who shot at the submarine and a pilot who saw it sink, says there is now no doubt the attack took place."

A congressional investigation known as The Roberts Commission was appointed by FDR to investigate the intelligence failure. It consisted of a four-member team. Two were members of the CFR. Like other congressional investigations puppeteered by the CFR, it was a whitewash. According to Perloff, Supreme Court justice Owen Roberts (a globalist), headed the commission. Perloff writes, "Two of the other four members were in the CFR." The commission absolved Washington of any blame and declared that the attack occurred due to "dereliction of duty" by Kimmel and Short.

"At the court-martials, attorneys for the defendants dug up some of Washington's secrets. The Roberts verdict was overturned: Kimmel was exonerated; Short received a small reprimand; and the onus of blame was fixed squarely on Washington," wrote Perloff. Apparently the Roosevelt administration repressed the results of the commission for reasons of "national security."

The Times described, "Despite testimony from Lieutenant Outerbridge and the seamen on the Condor ... [the account was] ignored by military and congressional tribunals investigating the attack on Pearl Harbor—including nine wartime commissions of inquiry—and the incident has since been omitted from historical accounts." Perloff added, "Incriminating memoranda in the files of the Navy and War department were destroyed. The court-martial findings were buried in a forty-volume government report on Pearl Harbor, and few Americans ever leaned the truth."

According to the *BBC*, the "attack on Pearl Harbor left 2,390 people dead, 1,178 wounded, 21 US ships heavily damaged and 323 aircraft damaged or destroyed." Smoot observed, "The council on Foreign Relations has heavy responsibility for the maneuvering which thus dragged America into World War II."

The CFR had what they needed to get us into WW II, now just a small amendment needed to be made to the constitution, so they could continue to legally fund the Nazis. Six days after the Pearl Harbor attack, President Roosevelt amended the Trading with the Enemy Act. "All this [funding] was done legally thanks to President Roosevelt," wrote Marrs. "This meant any kind of business transaction could be made legal with the approval of Roosevelt's Secretary of the Treasury, Henry Morgenthau, whose father helped found the Council on Foreign Relations," he stated.

September 11th

Entire books and documentaries are available on this subject. Most people have an understanding that things may not have occurred exactly as they have been told. Many believe they have been lied to in some way. And some believe that it may have been allowed to happen in order to facilitate a global war.

The reality of the situation is much worse. But rather than make claims which would require an entire book to support, I'll simply state that there are some very serious questions that need to be answered regarding not just 9/11, but the Okalahoma City Bombing.

To that I will add that the 9/11 Commission was chaired, vice chaired, and directed by members of the Council on Foreign Relations. I urge anyone interested in finding out what happened to view the documentaries, *9/11: Rise of the Police State*, by Alex Jones, *9/11: The Great Illusion*, by George Humphrey, and *Loose Change*.

Summary

So, not only did the U.S. crack the Japanese code and know what day the attack would occur, they had at least a one-hour advanced warning on the day the attacks took place. The CFR/Wall Street sacrificed over 2,000 people on December 7, 1941 for the sake of globalization. Then they used *their* history books to lie about it. If they really wanted to defeat the Nazis, they would not have continued to fund them up until 1944.

Chapter 12

Wall Street Funded Nazis

In his book, *Wall Street and the Rise of Hitler*, Professor Antony Sutton, author of nineteen books, provides a thoroughly documented account of the role played by Morgan, Rockefeller, General Electric Company, Standard Oil, National City Bank, Chase and Manhattan banks, Kuhn, Loeb and Company, General Motors, Ford, and other industrialists, in helping to finance the Nazis. To prove his point, Professor Sutton provides bank statements, letters from U.S. ambassadors, mainstream media sources, *Congressional Records*, excerpts from Congressional Investigations, and statements from the Nuremberg trials. Wall Street's funding of the Nazis is part of authentic history.

Professor Sutton wrote that "General Motors, Ford, General Electric, DuPont," and other "U.S. companies intimately involved with the development of Nazi Germany were ... controlled by the Wall Street elite," such as "the J.P. Morgan firm, the Rockefeller Chase Bank and to a lesser extent the Warburg Manhattan bank."

"The deal bringing Hitler into the government was cut at the home of banker Baron Kurt von Schroeder on January 4, 1933," wrote author Marrs. Other notable figures that are said to have appeared at this meeting include Council on Foreign Relations members John Foster Dulles, and Allen Dulles, of the New York law firm Sullivan and Cromwell, which represented the Schroeder bank. Allen Dulles would eventually become a member the Bilderbergers and director of the CIA.

"Max Warburg," stated Marrs, "a major German banker, and his brother Paul Warburg, who had been instrumental in establishing the Federal Reserve System in the United States, were directors of Interssen Gemeinschaft Farben or *I.G. Farben*, the giant German chemical firm that produced Zyklon B gas used in Nazi extermination camps."

"The financing for Adolph Hitler's rise to power was handled through the Warburg-controlled Mendelsohn Bank of Amsterdam and later by the J. Henry Schroeder Bank with branches in Frankfurt, London and New York," wrote Gary Allen. "Chief legal council to the J. Henry Schroeder Bank was the firm Sullivan and Cromwell whose senior partners included John Foster and Allen Dulles."

Author James Perloff concurs, and reveals the role that the Council on Foreign Relations played in aiding the Nazis. He states, "In 1939, on the eve of blitzkrieg, the Rockefellers' Standard Oil of New Jersey sold $20 million in aviation fuel to ... I.G. Farben [and] even had an American subsidiary called American I.G." Describing the CFR's connection to the Nazis, he lists the directors of the American I.G. as "ubiquitous Paul Warburg (CFR founder), Henry A. Metz (CFR founder), and Charles E. Mitchell, who joined the CFR in 1923..."

Other U.S. companies which contributed heavily to the Nazi war machine include Brown Brothers Harriman (BBH) and Union Banking Corporation (UBC), both of New York. Prescott Bush (grandfather of President George W. Bush) was a partner at BBH and director of UBC. UBC of New York, which was founded and chaired by E. Roland Harriman, is now confirmed to have been a *Nazi front company*.

In a story called, *Bush-Nazi Link Confirmed*, on October 10, 2003, *The New Hampshire Daily Gazette* announced, "After 60 years of inattention and even denial by the U.S. media, newly-uncovered government documents in The National Archives and Library of Congress reveal that Prescott Bush ... served as a business partner ... for the financial architect of the Nazi war machine from 1926 until 1942." A similar article appeared in the *London Guardian* on September 25, 2004, it was entitled, *How Bush's Grandfather Helped Hitler's Rise to Power.*

"Prescott Sheldon Bush ... father of the [former] President George [H. W.] Bush ... was a partner in the Wall Street firm of Brown Brothers Harriman for 40 years," wrote Professor Marrs. "It was Brown Brothers Harriman that helped to finance ... the 1917 communist revolution in Moscow and the rise of Hitler and Nazism, through money made possible by it and the affiliated Guaranty Trust Company." Prince Bernhard of the Netherlands who served in an intelligence unit for the German branch of Farben was also a member of the Nazi SS. Bernhard eventually became a founding member of the Bilderbergers.

In his book, *The Rockefeller File*, Gary Allen wrote, "The alliance between Nazi Germany and the Rockefellers is truly shocking." He explained, "Hitler's Luftwaffe ran on Standard petrol, and the Rockefellers were partners in I.G. Farben Industries, whose thousands of war products included the poison gas used in Nazi death camps." Professor Sutton added, "American I.G. Farben, General

Electric, Standard Oil of New Jersey, Ford, and other U.S. firms" were "directly responsible for bringing the Nazis to power."

"German bankers on the Farben Aufsichsrat (the supervisory Board of Directors) [of I.G. Farben] in the late 1920s included the Hamburg banker Max Warburg, whose brother Paul Warburg was a founder of the Federal Reserve System in the Unite States," revealed Professor Sutton. "Not coincidentally, Paul Warburg was also on the board of American I.G., Farben's wholly owned U.S. Subsidiary." So what this means is that Paul Warburg aided the Nazis while working for I.G. Farben in the U.S., while his brother helped them while working for the same company in Germany.

"The two largest tank producers in Hitler's Germany were Opel, a wholly owned subsidiary of General Motors (controlled by the J.P. Morgan firm), and the Ford A.G. subsidiary of the Ford Motor Company of Detroit," wrote Professor Sutton. "In brief," he affirmed, "American companies associated with the Morgan-Rockefeller international investment bankers ... were intimately related to the growth of Nazi Industry."

On March, 29, 2002, *The Guardian* ran an article entitled, *IBM Dealt Directly with Holocaust Organisers*. They wrote that, "Newly discovered documents from Hitler's Germany prove that the computer company IBM directly supplied the Nazis with technology which was used to help transport millions of people to their deaths in the concentration camps at Auschwitz and Treblinka." According to this mainstream publication, IBM supplied the Hollerith machines that determined when people would die, according to their weight, height, age, and other attributes.

Similarly, an article entitled, *How IBM Helped Automate the Nazi Death Machine in Poland*, appeared in *The Village Voice* on March 27, 2002, alleging that there was "a strategic business alliance between IBM and the Reich, beginning in the first days of the Hitler regime and continuing right through World War II." It continued, "Recently discovered Nazi documents ... make clear that IBM's alliance with the Third Reich went far beyond its German subsidiary. A key factor in the Holocaust in Poland was IBM technology provided directly through ... IBM New York, mainly to its headquarters at 590 Madison Avenue."

One excuse used by the industrialists that supported the Nazis, is that they had no idea what was going on. Unveiling this falsehood, Professor Sutton explained, "This financial and technical assistance is referred to as 'accidental' or due to the 'short-sightedness' of American businessmen." However, he stated "the evidence presented ... strongly suggests some degree of premeditation on the part of these American financiers." He further noted, "The general impression left with the reader by modern historians is that this American technical assistance was *accidental* and that American industrialists were innocent of wrongdoings."

The Kilgore Committee under the United States Senate charged with post World War II investigations concluded that "the United States accidentally played an important role in the technical arming of Germany. ... Germans were brought to Detroit [Ford Motor Company] to learn the techniques of specialized production of components, and of straight-line assembly. ... The techniques learned in Detroit were eventually used to construct the dive-bombing Stukas."

The report continued, "At a later period I.G. Farben representatives in this country enabled a stream of German engineers to visit not only plane plants but others of military importance, in which they learned a great deal that was eventually used against the United States."

Professor Sutton points out, "[The Kilgore Committee] makes it clear that I.G. Farben directors had precise knowledge of the Nazi concentration camps and the use of I.G. chemicals." To illustrate his point, he quotes a 1945 interrogation of I.G. Farben director von Schnitzler in which Schnitzler stated, "I was horrified" that the chemicals were being used in the camps but "kept it to myself because it was too terrible." Professor Sutton suggests, "not only was an influential sector of American business aware of the nature of Naziism, but for its own purposes aided Naziism wherever possible" and "profitable." He charges, "The pleas of innocence do not accord with the facts."

James Stewart Martin investigated the structure of the Nazi industry while working as Chief of the Economic Warfare Section for the Department of Justice. In his book, *All Honorable Men*, published in 1950, he asserts that American and British businessmen got themselves appointed to key positions in the post-war investigation. Martin concludes that this was a deliberate tactic to divert, stifle and muffle investigation of Nazi industrialists and to hide their own involvement. Regarding this matter Professor Sutton observed, "The evidence suggests there was a concerted effort not only to protect Nazi businessmen, but also to protect the collaborating elements of American and British business."

Referring to the Kilgore Committee in 1946, Professor Sutton wrote, "Accordingly, it is not at all difficult to visualize why Nazi industrialists were puzzled by 'investigations' and assumed at the end of the war that their Wall Street friends would bail them out and protect them from the wrath of those who had suffered." Apparently these Nazis were angry that they were even on trial and boasted that their "friends" in Wall Street would rescue them.

The Kilgore Committee in 1946 described the attitudes of these Nazi war criminals stating, "Their general attitude and expectation was that the war was over and we ought now to be assisting them in helping to get I.G. Farben and German industry back on its feet. Some of them have outwardly said that this questioning and investigation was ... of short duration, because as soon as things got a little settled they would expect their *friends* in the Untitled States and England to be coming over. Their friends, so they said, would put a stop to activities such as these investigations..."

The Control Council which was given the task of preparing directives for the arrest and detention of war criminals, moved into Germany after the war. It was headed by the Council on Foreign Relations. Professor Sutton observed, "At the end of World War II, Wall Street moved into Germany through the Control Council to protect their old cartel friends and limit the extent to which the denazification fervor would damage old business relationships." He said, "General Lucius Clay, the deputy military governor for Germany, appointed ... Banker William Draper [to] ... put his control team together from businessmen who had represented American business in pre-war Germany."

The CFR/Wall Street managed these investigations from the top. "So when we examine the Control Council for Germany," wrote Professor Sutton "we find that the head of the finance division was Louis Douglas, director of the Morgan-controlled General Motors," and the head of the "Economics Division was William Draper, a partner in the Dillon, Read Firm that had so much to do with building Nazi Germany in the first place." "All three men [Douglas, Clay, and Draper] were, not surprisingly ... members of the Council on Foreign Relations," he added.

"None of the Americans were ever prosecuted," wrote Perloff. "The story of American ties to German fascism has been avoided like the plaque by the major U.S. media." Professor Sutton declared, "After World War II the Tribunals set up to investigate Nazi war criminals were careful to censor any materials regarding Western assistance to Hitler."

"[At] the very core of Naziism," says Professor Sutton, we "find Wall Street, including Standard Oil of New Jersey and I.T.T., represented ... [up] to as late as 1944." Referring to Wall Street's role in funding the Nazis as being one part of a consolidation plan for world domination, he concludes, "This interplay of ideas and cooperation ... was only one facet of a vast and ambitious system of cooperation and international alliance for world control."

The professor quotes an Establishment insider; Georgetown professor Dr. Carroll Quigley, as saying that it was "nothing less than to create a world system of financial control, in private hands, able to dominate the political system of each country and the economy of the world as a whole."

Summary

According to this information, the Council on Foreign Relation, played an important role in funding the Nazis, and orchestrated the investigations to prevent their connections from being made public. This is a major historical lie. This evidence also suggests that Wall Street/CFR not only funded, but arguably, created the Nazi war machine.

Chapter 13

The Great Depression was Deliberately Created

Apparently congress was aware of the scheme of the international bankers and recognized the danger that the republic was in. Congressman Lindberg said in a *Congressional Record* dated, December 22, 1913, vol. 51, "This new law [the Federal Reserve Act] will create inflation whenever the trusts want inflation. It may not do so immediately, but ... if the trusts can get another period of inflation, they figure they can unload the stocks on the people at high prices during the excitement and them bring on a panic and buy them back at low prices... The people may not know it immediately, but the day of reckoning is only a few years removed."

"That day of reckoning, of course, came in 1929," said Perloff, "and the Federal Reserve has since created an endless series of booms and busts by the strategic tightening and relaxation of money and credit." Speaking about the historical disinformation regarding the crash, Perloff said, "Establishment historians present the '29 stock market crash as they do most events: an accident, evolved from erroneous policies, not from deliberate planning. We have all heard how foolish speculation bid stock prices high, but that the bubble finally burst, plunging brokers out of windows and America into the Depression."

"Having built the Federal Reserve as a tool to consolidate and control wealth, the international bankers were now ready to make a major killing," stated Allen. "Between 1923 and 1929," he described, "the Federal Reserve expanded (inflated) the money supply by sixty-two percent. Much of this new money was used to bid the stock market up to dizzying heights. At the same time that enormous amounts of credit money were being made available," continued Allen, "the mass media began to ballyhoo tales of instant riches to be made in the stock market. According to Ferdinand Lundberg: 'For profits to be made on these funds the public had to be induced to speculate, and it was so induced by misleading newspaper accounts, many of them bought and paid for by the brokers that operated the pools.'"

Perloff concurred, writing, "The Federal Reserve prompted the speculation by expanding the money supply a whopping sixty-two percent between 1923 and 1929. When the central bank became law in 1913, Congressman Charles Lindbergh had warned: 'From now on, depressions will be scientifically created.' Like two con men working a mark, the Fed made credit easy while Establishment newspapers hyped what riches could be made in the stock market." "Curtis Dall," he continued, "himself a syndicate manager for Lehman Brothers was on the floor of the New York Stock Exchange on the day of the Crash." Perloff quotes Dall as declaring, "Actually, it was the calculated 'shearing' of the public by the World-Money powers triggered by the planned sudden shortage of call money in the New York money market."

The "shearing," wrote Allen, caused a "despair [which] produced a willingness to accept a major expansion of government controls over the economy. ... In 1929, America was a long way from total government." He advised, "The next depression will be used as the excuse for complete socialist-fascist controls at home and the creation of a World Superstate internationally."

Congressman Louis McFadden, Chairman of the House Banking Committee, declared of the Depression, "It was not accidental. It was a carefully contrived occurrence." He warned, "The international bankers sought to bring about a condition of despair here so that they might emerge as rulers of us all." The Great Depression is another example of the *Problem-Reaction-Solution* formula.

"Plummeting stock prices ruined small investors, but not the top "insiders" on Wall Street," wrote Perloff. "Paul Warburg had issued a tip in March of 1929 that the crash was coming. Before it did, John D. Rockefeller, Bernard Baruch, Joseph P. Kennedy, and other money barons got out of the market. ... Early withdrawal from the market not only preserved the fortunes of these men," said Perloff, "it also enabled them to return later and buy up whole companies for a song."

"History shows that the Wall Street biggies came through very well indeed," wrote Alan B. Jones in his book, *How the World Really Works*. Quoting from G. Edward Griffin's book, *The Creature from Jekyll Island*, he added, "Virtually all of the inner club was rescued. There is no record of any member of the interlocking directorate between the Federal Reserve, the major New York banks, and their prime customers having been caught by surprise."

Jones quotes Herbert Hoover's description of the Secretary of the Treasury, Andrew Mellon's views, "Mr. Mellon had only one formula: 'Liquidate labor, liquidate stocks, liquidate the farmers, [and] liquidate real estate.'" [Mellon] said, "It will purge the rottenness out of the system. Values will be adjusted, and enterprising people will pick up the wrecks from less competent people."

"For those who knew the score," stated Allen, "a comment by Paul Warburg had provided the warning to sell. That signal came on March 9, 1929, when the *Financial Chronicle* quoted Warburg as giving this sound advice: 'If orgies of unrestricted speculation are permitted to spread too far ... the ultimate collapse is certain ... to bring about a general depression involving the whole country.'" "Sharpies [insiders] were later able to buy back these stocks at a ninety percent discount from their former highs," he declared.

"FDR is probably best remembered for the New Deal," stated Perloff. "Of courser, since a large portion of the work force was unemployed, there was not enough tax revenue to pay for these programs. So the government turned to its other source—borrowing. In effect, the international bankers, having created the Depression, now loaned America the cash to recover from it." He added, "Naturally, the interest on these loans would be borne on the backs of taxpayers for years to come."

Encyclopedia Britannica describes the Great Depression as the "Longest and most severe economic depression ever experienced," which "precipitated economic failures around the world" and triggered "major changes in the structure of the U.S. economy." "To think that the scientifically engineered Crash of '29 was an accident or the result of stupidity defies all logic," concluded Allen.

Summary

This evidence suggests that The Great Depression was artificially created so the larger Wall Street firms, which control the stock market, could eliminate competition and make profits out of lending America money to recover from it.

"Competition is a sin."
-John D. Rockefeller

Chapter 14

An Attempted Overthrow of the U.S. Government

In 1934 General Smedley D. Butler went public after he was approached by Wall Street and offered up to 300 million dollars to lead an army of 500,00 veterans equipped with munitions from Remington Arms Company to Washington D.C., in order to overthrow the U.S. Government by force. The plot was investigated by the McCormack-Dickstein committee, which found General Butler to be telling the truth.

During his 35 years of service in the Marine Corps, General Butler received the Congressional Medal of Honor twice. The Medal of Honor is the highest award for valor against an enemy force, which can be given to an individual serving in the Armed Forces of the United States.

Jules Archer's book, *The Plot to Seize the White House*, was written with the cooperation of senators, reporters, former Speaker of the House of Representatives John W. McCormack, former editors of the *Philadelphia Inquirer*, and the *Philadelphia Record*, and the immediate family of General Smedley Butler who provided military records, scrapbooks, memorabilia, recordings, etc. In his book Archer states, "School texts ... are uniquely silent about the powerful Americans who plotted to seize the White House with a private army, hold President Franklin D. Roosevelt prisoner, and get rid of him if he refused to serve as their puppet in a dictatorship they planned to impose and control."

The backing of the revolution reportedly came from the American Liberty League, which was headed by Morgan and DuPont. "Heavy contributors," says Archer, also included, "the Pitcairn family (Pittsburgh Plate Glass), Andrew W. Mellon Associates, Rockefeller Associates, E. F. Hutton Associates, William S. Knudsen (General Motors), and the Pew family (Sun Oil Associates)." The online encyclopedia, *Wikipedia*, added, "General Butler claimed that the American Liberty League was the primary means of funding the plot. The main backers were the DuPont family, as well as leaders of U.S. Steel, General Motors, Standard Oil, Chase National Bank, and Goodyear Tire and Rubber Company."

In his book, *Facts and Fascism*, investigative reporter, and author George Seldes wrote, "Most papers suppressed the whole story or threw it down by ridiculing it. Nor did the press later publish the McCormack-Dickstein report which stated that every charge Butler made ... had been proven true." Archer commented, "Press coverage of what was obviously a startling story of utmost importance to the security of the nation was largely one of distortion, suppression, and omission."

So, as expected, the mainstream press was called upon to discredit General Butler. "On November 22 the *Associated Press* struck a low blow at Butler by getting Mayor Fiorello LaGuardia, of New York, to express an opinion of the conspiracy based on what he had read about it in the press," wrote Archer. The *AP* ran this news item under the headline *Cocktail Putsch.* Explaining the contents of the article, Archer continued, "Mayor LaGuardia of New York laughingly described today the charges of General Smedley D. Butler that New York brokers suggested he lead an army of 500,000 ex-service men on Washington as 'a cocktail putsch.' The Mayor indicated he believed that some one at a party had suggested the idea to the ex-marine as a joke."

Commenting on a November 21, 1934, article on the front page of the *New York Times* entitled, *Gen. Butler Bares Fascist Plot To Seize Government by Force*, Archer wrote, "This was followed by a whole string of denials, or ridicule of the charges, by prominent people implicated. Extensive space was given to their attempts to brand Butler a liar or lunatic. Only at the tail of the story, buried inside the paper, did the Times wind up its account with a few brief paragraphs mentioning some of his allegations."

Continuing to quote denials from the article, he wrote, "It's a joke—a publicity stunt." "Thomas W. Lamont, partner in J. P. Morgan and Company, gave his comment: 'Perfect moonshine! Too unutterably ridiculous to comment upon!' J. P. Morgan himself, just back from Europe, had nothing to say. 'A fantasy!' scoffed Colonel Grayson M.-P. Murphy. 'I can't imagine how anyone could produce it or any sane person believe it. It is absolutely false so far as it relates to me and my firm.'" *Time Magazine* also ran a first-page story on December 3, 1934 that attempted to ridicule Butler under the headline *Plot Without Plotters*.

"If the press seemed overeager to emphasize denials of Butler's charges," wrote Archer, "the people of grass-roots America were far readier to believe the man who had exposed the plot. Letters of encouragement poured in from all over the country." Fortunately not all mainstream publications demonized General Butler. "Paul Comly

French broke the story [on November 20, 1934] in ... the *Philadelphia Record* and the *New York Post*," announced Archer.

Archer described French's article entitled, *$3,000,000 Bid for Fascist Army Bared*, writing, "Major General Smedley D. Butler revealed today that he has been asked by a group of wealthy New York brokers to lead a Fascist movement to set up a dictatorship in the United States. General Butler, ranking major general of the Marine Corps up to his retirement three years ago, told his story today at a secret session of the Congressional Committee on Un-American Activities."

Quoting Dickstein, Archer wrote, "General Butler's charges were too serious to be dropped without further investigation... He is a man of unquestioned sincerity and integrity. Furthermore, in my opinion, his statements were not denied or refuted. I think the matter should be gone into thoroughly and completely and I intend asking Congress for funds to make such an investigation. The country should know the full truth about these reputed overtures to General Butler. If there are individuals or interests who have ideas and plans such as he testified to, they should be dragged out into the open."

Unfortunately this was not the case. The committee, like all committees investigating the financial elite, was a whitewash. *Wikipedia* states that even though the committee, "did take the threat seriously and did verify that a fascist coup was indeed well past the planning stage, the Senate committee expired." This is similar to the Reece Committee's fate while investigating the Tax-exempt Foundations.

"Worst of all, no one involved in the plot had been prosecuted," wrote Archer. "MacGuire had denied essential parts of Butler's testimony, which the committee itself reported it had proved by documents, bank records, and letters," he declared. This is blatant perjury. So, he asked, "Did the department intend to file a criminal prosecution against MacGuire for perjury or involvement in the plot?" The answer of course is no.

The online encyclopedia noted, "Portions of Butler's story were corroborated," by "Veterans of Foreign Wars commander James E. Van Zandt," who attested that he too, had been approached by representatives of Wall Street to lead a "Fascist dictatorship" under the guise of a "Veterans Organization." General Butler's testimony was also supported by Captain Samuel Glazier—testifying under oath about a plot to install a dictatorship in the United States.

Of the committee's findings, the encyclopedia, states, "The Congressional Committee report confirmed Butler's testimony... In the last few weeks of the committee's official life it received evidence showing that certain persons had made an attempt to establish a fascist government in this country." It added, "There is no question that these attempts were discussed, were planned, and might have been placed in execution when and if the financial backers deemed it expedient."

The director of the ACLU, Roger Baldwin, stated, "The Congressional Committee investigating un-American activities has just reported that the Fascist plot to seize the government ... was proved; yet not a single participant will be prosecuted under the perfectly plain language of the federal conspiracy act making this a high crime."

Reportedly, on February 25, 1935 *Time Magazine* ran an article entitled, *Schnozzle, gimlet eye*: *Fascist to fascist* which stated that the HUAC found the plot to be true. Commenting on this, Archer stated, "In a tiny footnote at the bottom of the page, in five-point type that could barely be read, *Time* informed those of its readers with 20-20 vision, 'Also last week the House Committee on Un-American Activities purported to report that a two-month investigation had convinced it that General Butler's story of a Fascist march on Washington was alarmingly true.'"

In his broadcast over WCAU on February 17, 1935, Butler revealed that some of the most important portions of his testimony had been suppressed in the McCormack-Dickstein report to Congress. He said the committee had, "stopped dead in its tracks when it got near the top." "There is strong evidence to suggest that the conspirators may have been too important politically, socially, and economically to be brought to justice after their scheme had been exposed before the McCormack-Dickstein Committee of the House of Representatives," wrote Archer. He stated, "Powerful influences had obviously been brought to bear to cut short the hearings, stop subpoenas from being issued to all the important figures involved, and end the life of the committee."

Archer quotes retired Representative John W. McCormack, former Speaker of the House as saying, "There was no doubt that General Butler was telling the truth... We believed his testimony one hundred percent. He was a great, patriotic American in every respect." According McCormack, if General Butler had not exposed these plotters, Americans today could be living under a dictatorship.

When McCormack was asked by Archer why the Department of justice under Attorney General Homer Cummings failed to initiate criminal proceedings against the plotters. He answered, "The way I figure it ... we did our job in the committee by exposing the plot, and then it was up to the Department of Justice to do their job—to take it from there."

Not surprisingly, the alleged financial backing of this attempted overthrow includes a solid link to the CFR. "The founding president of the CFR was John W. Davis," wrote Perloff, "who was J.P. Morgan's personal attorney and a millionaire in his own right." In addition, Thomas W. Lamont was a founding member of the CFR. He also aided in the bloody Bolshevik Revolution. Perloff added, "Among the other Bolshevik abettors in the CFR's original membership were ... Morgan partner Thomas Lamont who helped persuade the British government to accept the New Soviet regime..." John J. Raskob, a director of the Liberty League, was also an early CFR member.

John L. Spivak, a reporter assigned to cover the committee hearings, called the story "one of the most fantastic plots in American history... What was behind the plot," said Spivak "was shrouded in a silence which has not been broken to this day. ... It would be regrettable if historians neglected this episode and future generations of Americans never learned of it."

Archer echoed, "It would seem time that school textbooks in America were revised to acknowledge our debt to the almost forgotten hero who thwarted the conspiracy to end democratic government in America."

Summary

According to this information, the CFR/Wall Street made an overt attempt to install a dictatorship by force. This plot was investigated by congress and found to be authentic. It is strikingly similar to the Communist Revolution which Wall Street is known to have funded, where an entire country was communized after a few major cities were seized.

"The Communists came to power by seizing a mere handful of key cities," wrote Allen. "In fact, practically the whole Bolshevik Revolution took place in one city—Petrograd. It was as if the whole United States became Communist because a communist-led mob seized Washington D.C."

Chapter 15

Commentary

The historical evidence you've just seen contains not just a couple of irrelevant details, that have for clarity sake been omitted, but major facts that have been deliberately left out of mainstream education. These facts would cause most people to view this world in a very different manner. Major historical lies have been widely distributed through the school system and mainstream publications. The lies have been relatively consistent, spanning multiple major historical events.

We've seen a pattern of the same financial elite (often the same families), lying, scheming, and instigating conflicts for their own profit and control. The media is their tool, used to attack their enemies, whitewash authentic history and keep people ignorant.

The controlling elite funded the savage Communist Revolution and the Nazi Holocaust, which resulted in the murder of millions. They built the dictatorship in Russia, which resulted in the brutal slaughter of millions of people during the Russian Holocaust. The same cabal allowed the attacks on Pearl Harbor to take place and deliberately created the Great Depression.

There was at least one blatant attempt to overthrow the U.S. Government by force in 1933-34, which congress confirmed to be genuine. Entire volumes could probably be written on their evil deeds. By control of the Tax-exempt Complex and other institutions, they control academia and have literally rewritten history.

Volume II Part I

to the Hidden Evil

Now that we understand who runs who runs this planet, it will be easier to see how such a secret policy can exist. It also becomes evident why money is no barrier. The reason of course is that they make the money. The *Hidden Evil*, is also called *Gang Stalking*, *Cause Stalking*, *Public Mobbing*, an *Electronic Concentration Camp*, *Covert War*, *Mental Rape*, and a variety of other names.(*)

It involves two major components. First, the use of Non-lethal Weapons (NLW), which includes Directed Energy Weapons (DEW). Second, heavily networked groups of plain-clothed citizens who encircle targeted people in public. This network appears to involve volunteers who are part of a state-sponsored *Neighborhood Watch* program, as well as individuals and groups of informants with no official connection to the state. As I'll demonstrate, mainstream news has indicated that one in twenty-four citizens will be recruited as informants. Some publications contend that the goal is the recruitment of one-third of the population.

Organizations from the federal level downward participate in the harassment, which includes local stores and restaurants. This program is not in the process of being created, it is already here. It is thoroughly embedded in most major communities. It is happening in all NATO countries. The program has been operational in some form since at least the early 80s, when families of people with direct or indirect connections to federal agencies were targeted.

Some people spent a significant about of time designing this program. Influential people have ensured its survival. It is also almost invisible and traceless. Some of the organizations I've found to be likely responsible for the Hidden Evil, contain people that are considered to be our world leaders. On the surface, they appear to be humanitarians.

The pretext for the existence of this program is presumably to identify and track potential threats to *national security* and to rid communities of undesirables. As I'll demonstrate, citizen informants have been used to stalk and harass people in other countries for the same reasons. Microwave weapons have also been used on dissidents.

U.S., German, and Russian intelligence agencies have all used these tactics. The Hidden Evil appears to be a culmination of them, which from what I have learned, has been fine-tuned with the help of Establishment Think Tanks.

Julianne McKinney is a former army intelligence officer and director of the Electronic Surveillance Project Association of National Security Alumni, which was an organization composed of former intelligence officers dedicated to exposing excesses by the U.S. Intelligence services. Her preliminary report, *Microwave Harassment and Mind-Control Experimentation* published in 1991, is a well-documented outline of these activities.

According to her interview on the *Republic Broadcasting Network* (RBN), these activities have increased significantly since the report was published. She kept this organization operational for four years using her personal funds. She has about 40 years experience with this subject and has talked with and interviewed hundreds of targeted individuals. I'll be referencing her report and her interview on the *RBN* throughout this book.

Dr. Reinhard Munzert, scientist, psychologist, and university lecturer has authored three books and over 100 articles. He began investigating this phenomenon with the aid of his students after learning he was targeted. His article, *Targeting the Human with Directed Energy Weapons*, which was the number one listing on Google for two years, will be referenced, as well as his interview on *RBN.*

I'll also be referencing the work of Leuren Moret, particularly her interview on *Vancouver Co-op Radio CFRO 102.7 FM*, on August 8th 2005. She is the former Environmental Commissioner of Berkeley California and the president of Scientists for Indigenous People. She has worked at two US nuclear weapons laboratories as a geoscientist and has contributed to a UN sub-commission investigating depleted uranium. She, Julianne and Dr. Munzert have been on the receiving end of this harassment.

In her article, *Uncovering the Truth About Depleted Uranium*, Moret speaks of her personal *Gang Stalking* experience, which she refers to it as *Mobbing*. She writes, "'Mobbing' is the purposeful and strategic institutional gathering of all information about an individual by using any method—legal or illegal. Contact for this purpose is made with neighbors, classmates, former partners, family members, former employers, teachers, church members, good friends and even your family tree..."

She continues, "Everything must be known about the support system around you that makes your life possible." This includes, "[the] assessment of strategic resources and future income (finances, mobility, cars, insurance, credit cards, bank safety deposit boxes, post office boxes, wills, health records, birth certificates, transcripts and photographs) is made without your knowledge. The purpose of mobbing is to drive the target individual out of the job force for the rest of their life; the bigger goal is to drive them to the ultimate self-destruction: suicide. The information gathered ... is used to take your life apart ... and to attempt to make you look crazy." She adds, "Your children are harassed, they come home with belongings missing and stories of teachers harassing them."

The psychological aspect of the group harassment which takes place in public is identical to *Mobbing*; it has the same group dynamics. For that reason I'll make references to Mobbing throughout this book. The *Mobbing* phenomenon has been well documented and studied in other countries for years. However, it is only now beginning to receive attention in North America.

Mobbing, also known as "psychological terror" involves the systematic harassment of an individual by a group of individuals in the workplace, and according to the authors of the book *Mobbing*, also in public. In the workplace, the group is often rallied by an authority such as Human Resources. It includes rumors, setups, and skits intended to humiliate the target and make them appear incompetent.

The documented mainstream exposure that Mobbing has received will help us understand how thousands of people in some major cities can be recruited to systematically attack individuals. The Mobbing I'm referring to is committed by multiple people using subtle but repeated harassment, which takes place in the workplace and is difficult to prove. It is used to kick a person out of the workplace or even workforce while leaving little trace. The logic is that it is less of a liability for the organization if you "voluntarily" leave than if they fire you. In addition, some organizations with sadistic intentions enjoy it as a form of *sport harassment* as well.

Although it would appear that the Hidden Evil is a new phenomenon, it is not. Tyrannical regimes throughout history have used local groups to neutralize dissent. *Group Stalking* is simply another manifestation of this pattern. In order to understand how an entire country could be complicit in the persecution of perceived undesirables, a study of German and Russian dictatorships is essential.

Therefore, evidence will be presented which describes how similar programs of persecution were used during these dictatorships, and how the U.S. and other NATO countries are heading in that direction. Finally, we'll draw upon the Milgram experiments, which will illustrate how most ordinary people will repeatedly torture innocent people under the direction of an authority figure.

Persecute
"To harass or punish in a manner designed to injure, grieve, or afflict; specifically: to cause to suffer..."
-Merriam-Webster's Dictionary

Persecution
"Persecution is the systematic mistreatment of an individual/group by another group. The most common forms are religious persecution, ethnic persecution, and political persecution."
-Wikipedia Encyclopedia

Endnotes
* These names are taken from a combination of web sites listed in the sources.

Chapter 17

Tactics

It has been said that when you're first targeted, they hit you hard in multiple directions. Presumably, this is done so you'll collapse after experiencing recurrent emotional violence, which is compounded after you've realized that all normal support structures and avenues of grievance have been removed. There appears to be multiple levels of intensity of this program that people can receive. The two main parts of the program are group stalking and Non-lethal Weapons (NLW). Both are violent attacks. And some are apparently hit hard with both.

People tend to associate the word *harassment* with acts of temporary aggravation or pestering. Therefore, it may not by the proper word to describe the never-ending acts of covert violence inflicted upon targeted people, which amounts to torture and murder. Some of the tactics used in these programs are borderline-subliminal, which is why they are so difficult to detect, explain, and defend against. It is my goal to make the unconscious, conscious. Many of these tactics are a type of *mind-control*.

Most of these tactics have been used in former *Cointelpro* operations by the United States, Russia, and Germany against their civilian population. The Hidden Evil's genesis is obviously derived from known tactics such as bogus investigations, surveillance, smear campaigns, the use of noise, thefts, break-ins, character assassination, staged accidents, framings, sabotage and vandalism, mail interference, blacklisting, citizen informant squads for overt surveillance (Gang Stalking), poisoning and drugging, incarceration into prisons and mental institutions, and electromagnetic weaponry.

It is a historical fact that the U.S. has officially targeted its civilian population on multiple occasions. Programs such as MKULTRA and Cointelpro can't be denied. "The methods reportedly employed in these harassment campaigns," explained McKinney, "bear a striking resemblance to those attributed to the CIA and FBI during Operations MKULTRA, MHCHAOS, and COINTELPRO."

In Germany these methods were contained in two programs, *Zersetzung* and *Directive Perceptions*. They were basically German versions of Cointelpro. Zersetzung was developed to destroy a targeted person's inner-self. During an interview on the *Republic Broadcasting Network*, Dr. Munzert, a German target, drew a parallel between the current program, and what occurred in East Germany. He explains, "In

Germany this is called *Zersetzung* and this means the dissolution or disintegration of the targeted person."

Writer, international lawyer, and filmmaker Anna Funder won the Four Samuel Johnson Prize from the *BBC* in 2004 for her non-fiction book, *Stasiland.* In it, she wrote, "The German word *Zersetzung* is harsh, and has no direct English equivalent." She described, "Zersetzung, as a concept, involves the annihilation of the inner-self." It calls for, "[The] targeted spreading of rumours about particular persons with the aid of anonymous [means] ... making compromising situations for them by creating confusion over the facts ... [and] the engendering of hysterical and depressive behaviors in the target person."

"Directive Perceptions," she continued, was another program designed to "develop apathy (in the [targeted] subject) ... to achieve a situation in which his conflicts, whether of a social, personal, career, health or political kind are irreversible ... to give rise to fears in him ... to develop/create disappointments ... to restrict his talents or capabilities ... to reduce his capacity to act and ... to harness dissentions and contradictions around him..." These directives were used by the Ministry for State Security (MfS) against threats to state security. In other words, they were massively used on the population to maintain a dictatorship.

These programs were designed to secretly destroy a target's personal and professional life, ruin them financially, and prevent them from reaching their potential, with the intention to produce irreversible depressive and panic-stricken behaviors. What the Hidden Evil appears to be, is a combination of various Cointelpro operations used by oppressive regimes on their civilian population. It has been refined and deployed globally. As I'll demonstrate, it is likely an indication of the creation of a worldwide dictatorship.

Hidden in Plain View

Many of us have experience with books where the goal is to find the hidden picture contained in the landscape. Or the artwork, where if you look intently, you will eventually see a three-dimensional picture superimposed on a flat background. At first, most of us had some trouble detecting these hidden pictures, but with practice we usually see them. Once the detection mechanism is developed, these hidden objects become clearly visible with little effort. So too with the Hidden

Evil. Once you know what to look for, these borderline-subliminal attacks are quite detectable.(*)

So, rather than using a blunt overt attack against a fully functional person, they usually attack covertly, and conceal their harassment by using existing events that occurs naturally. There is usually cover for the harassment as well. However, it is probably a statistical rarity for some of these *staged events* to occur as often as they do. They conceal much of their harassment using what appears to be a simple formula. From what I've learned, the basic formula is as follows:

- *Frequency:* Describes how often an event occurs. It also pertains to the number of acts within a single event.
- *Duration:* Pertains to the length of a single event. It also pertains to the non-stop nature of the harassment in general.
- *Intensity:* The amplification of acts such as sound, sight, crowding, etc. within an event.

A hypothetical example of a car door-slamming event would include this: *One neighbor arrives and two others leave. As they're tending to their vehicles there is the repeated opening and shutting doors and trunks, alarms going off, and beeping from alarms being turned on. These disturbances emanate from areas surrounding your house. This may happen simultaneously, or they may be strung together, one right after the other.*

Synopsis

The event lasted for 5 minutes; longer than normal (Duration). In addition to length of the event being extended, the number of times each act occurred within the event was increased (Frequency). Ex: Multiple, repeated doors/trunk slamming from cars, even if a single individual is at the vehicle. This event also happens many times throughout the day (Frequency). Each individual act of door/trunk shutting is amplified by deliberate slamming to produce a louder than normal noise (Intensity).

The event was louder, contained more activity (acts), and lasted longer than normal to produce a covert attack. Even if a single individual is arriving/leaving, standard practice seems to be multiple slams (trunk, back door, driver's door). This may be synchronized with your activity, such as your arrival or departure. The event may also be a part of a *Noise Campaign* consisting of an alteration of other types of noise. This basic formula is used with many of the tactics outlined in this chapter. The *frequency and duration* formula is also used in Mobbing. In addition to the overuse of an individual tactic, these tactics are combined and used in a round-the-clock fashion, which amplifies their overall potency.

Basic Protocol

There is a basic protocol that they begin with which is consistent in all NATO nations. It begins with the surveillance of targets, monitoring of their private lives, and entry into their homes (break-ins). This is done so their personality traits can be cataloged. "There is a basic protocol that the perpetrators begin with," states McKinney, "but the *TI* [Targeted Individual] contributes to the modification." After they are singled out for preliminary stages of harassment, Gang Stalking ensues, which McKinney describes as part of a "softening-up process." After a period of overt surveillance (Gang Stalking), NLW are introduced. The NLW harassment gradually increases to extreme conditions. This pattern has unfolded consistently in all NATO countries.

Dr. Munzert speaks of basically the same pattern, which he describes as a "double-folded strategy." "It is usually the same procedure," he announced, "but with individual variations." One part includes the victims being "attacked with microwave weapons," and "the other part of the strategy" he says, is to portray the targeted people as "mad [insane]." He explained the effectiveness of this approach as "unbeatable," and reveals that this is essentially what the Stasi did to their targets. Part of this protocol appears to include elements of *Neuro-linguistic Programming* (NLP), which is a type of mind-control used by behaviorists to affect change. NLP will be covered in more detail shortly.

Some of the tactics below have been called *Street Theatre, Harassment Skits,* or *Staged Events.* They are planned harassment skits, such as blocking, or swarming. They also include informants who surround targets and have conversations intended to be overheard, which contain information about the target's personal life. Presumably, citizen informants are told that this is necessary to let targeted people know they're being watched. This may happen in any public place. This is not a complete list of tactics but it includes some of the more common ones reported.

Some of these tactics will sound insane because they're deliberately designed to make someone appear as though they're suffering from a mental disorder. They were definitely created by experts in the behavioral sciences. In addition, people may be emotionally drained, and unable to properly identify or explain what's happening to them. According to the DOJ, mental tactics designed to cause psychological harm must last for months or years before they constitute torture. The Hidden Evil fits this description. So keep in mind, these are Psychological Warfare tactics which are intended to drive people *crazy*.

Mental Health System

The mental health system is apparently being used worldwide to discredit targeted people who complain. According to the book, *Journey Into Madness: The True Story of Secret CIA Mind Control and Medical Abuse*, by Gordon Thomas, countries around the world, including the U.S., have used doctors to help abuse and discredit people, often for political reasons. It is also a documented fact that there was collaboration between doctors and the military when experiments were carried out in prisons, hospitals, and universities on unwitting people in North America.

McKinney states that medical doctors, as well as psychiatrist and psychologists appear to be providing cover for this program. This is similar to what was done in Russia, where enemies of the dictatorship would be thrown into mental institutions and drugged beyond recovery. "The APA's refusal to acknowledge the impact of terrorization upon the human psyche, even given the publicity stemming from the Church Committee's findings in 1975, raises serious questions about the validity of psychiatry as a profession in this country," declared McKinney, "not to mention, the APA's ethical intent, in the long term."

The *DSM* is a diagnostic manual for identifying mental disorders. The first edition of the DSM was released in 1952 at a time when the APA was under the control of Dr. Ewen Cameron, who would commit brutal government-sponsored torture under the MKULTRA program. Dr. Rauni Leena Kilde, former Chief Medical Officer of Finland, wrote an article entitled, *Microwave Mind Control: Modern Torture and Control Mechanisms Eliminating Human Rights and Privacy*. In it she described "*The Psychiatric Diagnostic Statistical Manual* (DSM) for mental disorders," as a "brilliant cover up operation in 18 languages to hide the atrocities of military and intelligence agencies' actions towards their targets."

Dr. Kilde says the manual "lists all mind control actions as signs of paranoid schizophrenia," and that "all medical schools teach their students that the person is paranoid, ESPECIALLY if he believes intelligence agencies are behind it all." Finally she proclaims, "Never is the medical profession told that these are routine actions all over the world by intelligence agencies against their targets."

Dr. Munzert said that doctors, "first think of paranoia and schizophrenia" if someone complains of this program. He cautions that because of this, "victims" could end up in a "lunatic asylum." Annie Earle, a Licensed Clinical Social Worker, and Board Certified Diplomat in her area of expertise, has over 25 years of psychotherapy experience as an independent practitioner. When referring to the DSM during an interview on the *Republic Broadcasting Network*, she stated, "It's called the Diagnostic Statistical Manual ... [and all] psychotherapists, regardless of specialty are required to give their patients a diagnosis from this manual, and in order to get the diagnosis you have to fulfill certain criteria that is clearly spelled out in the *DSM*."

Earle noticed a pattern unfolding when some of the people who came to her who did not fit the standard criteria. "The patients that started coming to me really did not fit any of the standard classifications," she explained. "For example, they might be having what some people consider a psychotic or schizophrenic breakdown in that they might report what seem to be hallucinations, but they did not have any of the other criteria that goes along with schizophrenia. There are many criteria that one is required to ... [detect] in making the diagnosis of schizophrenia, and these people did not meet all that criteria." Apparently Earle became targeted herself while studying this program.

Interestingly, the April/May 1996 issue of *Nexus Magazine* contained an article entitled, *How The NSA Harasses Thousands Of Law Abiding Americans Daily By The Usage Of Remote Neural Monitoring* (RNM) written by former NSA worker John St. Clair Akwei. It read, "NSA DOMINT [Domestic Intelligence] has the ability to covertly assassinate U.S. citizens or run covert psychological control operations to cause subjects to be diagnosed with ill mental health."

The *DSM* appears to have been created with a loophole which allows government-sponsored harassment programs to exist. This was evidently done so that people will appear ill when explaining this to a mental health professional. Adding to this confusion, the program itself is designed to mimic mental illness when explained to those "not-in-the-know." Mental health professionals may also participate in this program, wittingly or unwittingly by labeling targets as mentally ill. It is one of several layers of protection used to help keep this program operating in secrecy. It's easy to see how some mental health professionals might "jump to conclusions."

Surveillance

Surveillance is apparently part of an ongoing bogus investigation that never ends. The investigation may be done for reasons of "national security." Similarly, German and Russian security forces would often target the civilian population for the same reason. After a person is singled out, dwellings surrounding the target will be sublet and used as bases of operation and training.

If the target lives in a house, the base will usually be the closest building to the house. If the target lives in an apartment, then the base will be in one or more of the surrounding apartments. There will be unusual patterns of occupancy with lots of people coming and going into these bases. Vehicles and license plates used by the occupants of the bases may be periodically changed. According to McKinney and Moret, the people who move into these bases have been linked to the DOJ, CIA, and other federal agencies, as well as Universities, and government defense contractors.

"In order to target someone it requires that that person be put under surveillance," proclaimed McKinney, "so that their personality traits, their capacity to inter-relate with people, their capacity for corruption or non-corruption (that seems to be a critical point), and even their religion factors into it." The information obtained by surveillance is used for no ethical reason, "other than to establish a harassment protocol, which will follow that targeted individual for the rest of his or her life," said McKinney. Moret said that neighbors, friends and family are then co-opted into the harassment.

This surveillance is apparently done using very sophisticated equipment which can locate targeted people through walls. This equipment seems to be used in combination with Directed Energy Weapons and other "Non-lethals." According to McKinney, miniature microphones and cameras may be planted in your home, or through-the-wall infrared devices may be used to monitor your movement. The Life Assessment Detector System (LADS) developed at Georgia Tech was apparently created to rescue people trapped in rubble, avalanches, or other disasters. It has a range of about 135 feet can and can detect heartbeat and respiration. It can also be configured to track a specific person's movement inside a building.

SoldierVision A1, a radar flashlight, has a standoff distance of 30 feet and can detect motion behind a wall at 30 feet as well. These flashlights have been steadily researched using military and university testing labs since the 1990s. Some of these flashlights display images in 3D format, can detect respiration, and can see through, wood, brick, plasterboard, glass and concrete. More than likely, the surveillance technology which is now being used on targets worldwide is much more advanced and not publicly available.

Character Assassination

Targets usually experience character assassination which is done to destroy their personal and professional life. Half-truths and lies are often strategically spread into their lives, and may be the basis for a bogus investigation. The rumors circulated about people are apparently not petty, but shocking and repulsive. They are CIA or Stasi-type life-damaging lies. Some of the people spreading these lies may be acquaintances and/or family members that have been recruited (lied to, intimidated and blackmailed) into becoming informants. More on the recruitment of family and friends will be covered in the *Informants* and *Motivational Factors* chapters.

U.S. Federal law enforcement has used these tactics to destroy people they either wanted to recruit as informants or those they knew they couldn't recruit and just wanted to destroy. According to former CIA psychologist James Keehner, "It was planned destructiveness. First, you'd check to see if you could destroy a man's marriage. If you could, then that would be enough to put a lot of stress on that individual, to break him down. Then you might start a rumor campaign against him. Harass him constantly." Constant harassment, planned destructiveness, character assassination, destroying relationships, all done in concert with the goal of breaking someone down are a standard part of this program.

Reportedly, the usual rumors spread about a person are that they are a pedophile, prostitute, terrorist, racist, anti-government, mentally ill, drug addict, drug dealer, or a threat to "national security." They are often spread to neighbors, friends, family, employers, etc. A major effort is spent to separate a person from their friends and families.

Apparently these rumor campaigns aid in recruitment. Dr. Rauni Kilde wrote, "Deception is the name of the game, so recruits are told untrue sinister stories of their victims to keep them motivated." According to *CNN*, the Stasi used the exact same tactics. They stated, "The agency was authorized to conduct secret smear campaigns against anyone it judged to be a threat..." The Russian KGB too, with its massive network of informants, would slander their internal targets.

These rumors may be propagated by multiple people (informants) behind a target's back. Apparently, not letting a targeted person become aware of these rumors is essential to their effectiveness. Victor Santoro noted in his book, *Gaslighting: How to Drive Your Enemies Crazy*, that "The essence of defaming your target with rumors is it that it not get back to him."

Some targeted people believe that rumors are being spread that they are pedophiles. The organizers realize that labeling someone a pedophile will be highly damaging and produce optimum leverage for gaining community support. It is probably one of the worst rumors you could launch against someone. The Stasi would destroy the character of their targets by labeling them as pedophiles as well. There have been multiple magazine and newspaper reports of pedophiles being harassed by neighbors, and vigilante groups to the point where they are driven out of communities and forced out of work. Some have been driven out of multiple states.

An article in the *Toronto Star* entitled, *Vigilantes Versus Pedophiles* on August 8, 2004 outlined how one citizen's group called, Perverted-Justice, would start campaigns to expose people that they have concluded were potential pedophiles. This illustrates the mind-set of some groups of people who have reason to believe a person is a pedophile. The PJ group is run by Xavier Von Erck (real name Phillip John Eide), at www.perverted-justice.com.

Upon visiting a site called Corrupted-Justice at www.corrupted-justice.com, I discovered that multiple members of Perverted Justice have endured trauma such as incest or other forms of sexual abuse and are suffering from a variety of mental disorders. The Corrupted-Justice site obtained this information from member profiles which are publicly available, and located at the Perverted-Justice website. And according to CJ, Eide is on record saying that more than 40% of his followers have endured this type of sexual abuse.

The CJ site also has some very interesting message board chats (publicly available) by PJ members that are worth a look. Apparently the CJ people have documented proof that members of PJ have hacked into their web server. The CJ people are for the full prosecution of potential sex offenders by using law enforcement equipped with the checks and balances built into the legal system. The CJ group basically says that although the PJ system appears to be beneficial, it is heavily corrupted.

Most PJ "busts" do not include law enforcement. According to the article in the *Toronto Star* and the CJ website, the volunteers at PJ instead begin a relentless campaign to expose people they believe are pedophiles by calling friends, neighbors, children, places of employment and ruining personal and professional relationships. This is similar to what targeted people experience, as indicated by the behavior of their friends, family and co-workers. But they're never told about it. PJ is listed as a potential destructive cult by the Rick A. Ross Institute for the Study of Destructive Cults, www.Rickross.com.

It appears that some PJ members have a psychological need to feel powerful and meet that need by displaying to the public and their targets how they can destroy a life. I believe that some of these volunteers are ill and have latent sadistic intentions, despite their *patriotic* motives. Apparently some of them receive a sick type of enjoyment/empowerment by ruining the lives of people, as evident by their chats.

Character assassination is standard. My research leads me to conclude that it is not beyond these people to masquerade as "concerned citizens" and call local officials to register destructive complaints about a targeted person. I'm fairly certain they do this. This type of "community service" can be very damaging, especially when the complaints originate from multiple, seemingly unconnected sources.

Sensitivity Programs (NLP)

Some of the sensitivity tactics used by these groups are borderline-subliminal attacks designed to artificially create phobias. They are apparently based on *Neuro-Linguistic Programming* (NLP). You can think of *NLP* as a very powerful tool that can be used to produce rapid, profound change. The parts of NLP that groups use are *anchors* and *triggers*.

The goal of a these sensitivity programs appears to be to condition people with damaging emotions which are linked to *triggers* such as objects, colors, movements, and sounds. Once this is done, a target can be covertly injured openly in public. This may happen with or without the target's conscious awareness. As I'll demonstrate, this is brutally violent.

"There is a basic protocol that the perpetrators begin with, but the TI contributes to the modification," explained McKinney. She described that a "pavalonian conditioning" program, is used to get targeted people to "respond emotionally to a particular trigger." These negative emotions are then "built into the protocol." "It's an ongoing process," declared McKinney. Moret agrees saying, "They will stalk the target for a while ... to condition them," and make them "confused and frightened."

Self-help gurus use these programs to create positive emotional states and *anchor* them to a movement, a sound or an object. Then the sound, movement, or object becomes the *trigger* that will invoke the emotion. Although this sounds complex it is pretty simple in practice. It is done by creating a peak emotional state, and then while in that state, you *anchor* it, that is, do something repeatedly. This effectively *anchors* the emotional state to whatever was repeatedly done.

"A stimulus which is linked to and triggers a physiological state is called an anchor in NLP," stated O'Connor and Seymour in their book, *Introducing NLP*. They ask, "How are anchors created?" "First by repetition... Secondly, and much more important, anchors can be set in a single instance if the emotion is strong and the timing is right." This process of creating anchors has also been called *Emotional Transference*.

We're unconsciously creating anchors in our environment all the time with people, music, places, and objects. These stalking groups are obviously led by people knowledgeable in the behavioral sciences. The groups are used to create negative emotional states in targeted people, such as fear and anxiety, and anchor them to common objects. This is the deliberate creation of a phobia!

O'Connor and Seymour declare "An external stimulus can trigger a very powerful negative state." "This is the realm of phobias," they warn. This is nothing less than using violence to create an injury and then deliberately irritating it. "Targets are constantly monitored and if a target responds emotionally to a particular trigger, that will be built into the protocol," says McKinney.

Organizations known to have studied NLP include military intelligence agencies, CIA, FBI, and other state bureaus of investigation. According to an article called, *Non-Lethality*, which appeared in June 1993 issue of *Lobster Magazine*, in the early 1980s, Dr. John Alexander taught NLP to "selected general officers and Senior Executive Service members" as a "set of techniques to modify behavior patterns." Some of the references used in this article were taken from

Dr. Alexander's book, *The Warrior's Edge*, which was co-authored by Janet Morris, who, according to the article, was a student of the Silva course in advanced mind-control. In the S*tructure* chapter, I'll provide evidence that Dr. Alexander and Janet Morris have contributed to the creation of this program on behalf of the *Think Tanks*, as part of a political agenda.

In their book, Dr. Alexander and Janet Morris wrote, "In 1983, the NLP training group, along with John Alexander, was engaged to teach these skills to several members of Congress, including Al Gore and Tom Downey, under the auspices of Congressmen Clearing House on the Future, a bipartisan activity established to provide information to congressmen when they request it." He continued, "One organization that played a major role in the dissemination of NLP skills was the U.S. Army Intelligence and Security Command (INSCOM). ... Unlike the rest of the U.S. Army, INSCOM does not differentiate between wartime and peacetime roles."

According to his Bio, available at www.platinumstudios.com, Dr. Alexander has worked at "the highest levels of government," such as "the White House Staff, National Security Council, Members of Congress, Director of Central Intelligence, and senior defense officials." Morris' Bio, available at www.m2tech.us, is similar. Some of the organizations she's worked with include: the White House Office of Science and Technology Policy, Office of the Secretary of Defense, the Marine Corps Warfighting Lab, etc.

"Anchoring," stated Morris and Alexander, "is based on the neurophysiological assumption that patterns of behavior can be installed, then reactivated whenever a similar situation is encountered or created. ... Knowing this, you can intentionally install anchors to return your target to specific emotional states. The first step is [to] create the desired emotional state in the individual."

There has been some concern in the mental health field over the potential misuse of NLP. Unfortunately, like other tools, such as guns, they can be misused in the wrong hands. Andreas and Faulkner warned of the "possible uses and misuses of this technology," in their book, *NLP: The New Technology of Achievement*. They wrote, "We recognize the incredible power," of NLP and "recommend you exercise caution..." The deliberate infliction of emotional pain is an *act of violence*. The use of NLP in this manner is similar to a physical attack such as a punch or kick, only it leaves no visible injury.

Here's an example of how groups use NLP to make you aware that you're being stalked:

One morning you're walking out to your car and a man walks in front of you, stares directly at you in a hostile manner while repeatedly clicking his pen the whole time. You feel a little uneasy and say to yourself, "Boy that was weird, what was his problem?"

Then two hours later you're leaving a shopping mall on a day that it was not particularly crowded. As you walk out to your car, an old lady approaches you on foot. As she almost collides with you, she locks her eyes onto you with a hostile demeanor, while clicking a pen the whole time. Then maybe you tell yourself something like this. "There couldn't be a relation between those events, probably just a coincidence. But they both stared at me in such a mean way while hammering their pens..."

Later in the afternoon, you're on your way home. You're at a stoplight and out of the corner of your eye you see a passenger in the car to your left. You hear a slight noise, and glance over. You see a man glaring at you with a grin on his face. His arm is hanging out the window and he is holding a pen, which he is repeatedly clicking. Now imagine that happening for a month. Then a few days later you're on a sidewalk with a friend or family member. A man walks by, doesn't look at either one of you, but clicks his pen a couple of times. You feel anxious and afraid.

You may not even know you've been sensitized. One reason people may not realize that they're targeted or can't remember exactly when it began, is because these groups may slowly and gradually increase the harassment over time. But if you did recognize it as an attack, what would you tell your friend or family member?

This is what happened: The collisions, blatant hostile staring, and foul grin created the negative emotions and *anchored* them to the pen, which has now become the *trigger*. That is an example of a sensitivity program. Now imagine yourself being sensitized to multiple objects and sounds, and each of them creating pain each time you see or hear them.

Targets around the world have witnessed this being done with sounds, gestures, cell phones, laptops, pens, cars, watches, clothing, symbols, colors, and other items. This type of attack does require some maintenance and will loose its potency unless it is reinforced. So groups will reinforce these anchors and triggers with an occasional blitz attack from time to time.

The example above is a simple tactic which groups can use to let you know that you're being harassed in public. But there are many other attacks which NLP can be used for. For instance, you have recently experienced an extremely traumatizing event (which may have been facilitated by a faction of this network), such as an accident, death of a family member or pet, or some type of a brutal assault. Those who have you under surveillance know that there is a color, object, or sound that you've linked (anchored) massive pain to. Then shortly afterward, you are stalked by multiple people who carry that object, wear that color, or utter that phrase. This brings to the surface all the emotional pain you suffered during that experience. Imagine this happening again and again, every time you go outside. Now think what it would be like if your family and friends started to participate.

Groups will take an object you've been sensitized to and link it to another object. The idea appears to be to keep expanding the amount of objects you associate fear, anxiety, anger, or shame with. For instance, one of your neighbors has made it clear that they're participating in the harassment. Now, you know they're participating, and they know you know. So they may try to sensitize you to another stimulus such as a car alarm.

Because you've already associated them with massive pain, they can extend their harassment to a sound by turning their car alarm on/off in rapid succession over a short period of time, multiple times throughout the day. After they've done this for a couple of days and you've been sensitized to this sound, they can reduce their use of this specific tactic and just use the occasional maintenance sounds to inflict pain.

Now instead of repeatedly turning it on/off ten times, they only do it two or three times in a row, just to let you know you're being attacked. Even though it's only done a couple times in a row, you know why they're doing it and you may feel anger, fear, anxiety or other negative emotions. You may also feel frustrated at the prospect of trying to explain this harassment to another individual.

Anyone else observing this might think it's a little strange that someone would turn their alarm on/off a few times but they'd probably write it off as an isolated strange incident. But because they have not had your experience with that sound, and are not aware that it is a small part of a much larger harassment program, it would be difficult to explain that those beeps were attacks.

In NLP, the process of copying an emotional state from an existing trigger to a new trigger is called *chaining*. O'Connor and Seymour state, "Anchors can be chained so that one leads to another. Each anchor provides a link on the chain and triggers the next one, just as the electrical impulse flows from nerve to nerve in our body."

After a person has been sensitized to a color or object, the article can become a unification symbol for the group, much like a uniform. For instance, after a person has been sensitized to the color red, she is surrounded by people wearing red cloths in public. Furthermore, this is an *adaptable uniform* because it can be changed in less than five minutes. If a person who realizes that they're being Gang Stalked has been sensitized to red, then the organizers of these groups can simply have them blitzed by a horde of citizens wearing the color blue. The harassment has now been *chained* to this color. Most likely, these uniforms also help promote group cohesiveness, and may foster feelings of empowerment among stalkers.

Space Invasion (Crowding)

Space invasion includes blocking, cut-offs, and swarming or what some TIs refer to as, *crowding*. Targets may be encircled by people wearing colors or holding objects that they've been sensitized to. Prolonged crowding can have an extremely negative effect on your mental/emotional health. The people who designed this harassment program were obviously well aware of this. While some cultures may be more immune to crowding than others, even people from parts of the world that have been historically crowded react stressfully to crowding. People who have not been invited into your personal space but deliberately violate it are attacking you.

Body language is more accurate and reliable than verbal communication. Most of this non-verbal communication is unconscious. The study of this language has been called *Social Kinesics* (the study of body communication). There are two components to this communication: One contains movements, gestures, postures, the other is spatial relationships. The study of communication using distance/space is called *Proxemics*.

For most North Americans the intimate zone is 0-18 inches and personal space is 1½-4 feet. This type of communication is used to harass a person. For instance, while in stores, restaurants and public places that are not busy, a target will have people invading their personal space. This happens even if there are only two customers in a store. If a targeted person makes a move from one place to another, several people may suddenly appear and jump out in front of him from around corners and isles.

The deliberate space invasion use by these people is designed to look like normal cut-offs and collisions that we all experience from time to time. It can happen with vehicles, on foot, or a combination. This type of space invasion is obviously intended to startle people and make them tense. It can be thought of as a *virtual slap*. They are used in blind areas such as corners, hallways, restrooms, or intersections where targets have people or cars cutting them off, or almost hitting them in a calculated manner. This happens in stores, buildings and on the street, with people and vehicles.

With some artificially induced cut-offs, the perpetrator is seen at a distance. The target and the perpetrator(s) are heading toward the same focal point which is used to maximize crowding. The point may be a corner, a very small walkway, a thin passageway, or an obstruction on the sidewalk such as a telephone pole or a tree. The informants adjust their timing so that they meet the target at the exact point where there is the least amount of space for all parties to pass, thus maximizing the invasion of the target's space. Apparently the idea is to get the target's attention and make them uncomfortable. When standing in a checkout-line, targets may routinely have group members standing a foot or closer behind them. And rather than wait for a target to leave, they may crowd them as they reach to the side or in front of the target to put their items on the counter.

If a target is outdoors in a populated area, he or she may see a wave of people move in the direction that they're headed in order to swarm them. If a target moves, then stops and waits for a few moments, they may see that the activity will settle back to normal again after the horde passes. This test can be run indoors and outdoors as well. Sometimes the horde may blatantly adjust their timing, or linger for a while until the target begins to move again, then they'll continue. These crowding and blocking tactics amount to an obstacle course that follows people wherever they go.(*1)

They will also block targets, both on foot and with vehicles. For instance, as a target leaves a parking lot, a vehicle or person will be in the way, usually for a few moments. These things happen to everyone, once in a while, but for targets, it happens consistently, regardless of the time of day or how busy the area is.

This space invasion happens on days/times that there are normally not many people. It happens in stores while making purchases, restaurants, businesses, where targets have people frequently cutting them off, invading their space, using sensitivity programs, or other harassment tactics. If there is one customer in the store, they will probably be invading the target's space. For instance, a TI is indoors when it is not crowded, but witnesses multiple people who, instead of taking the shortest path to their destination, go double or triple the distance just so they can brush by them or cut them off.

Here is one personal example which took place at a YMCA in Boston, shortly after I realized I was being Gang Stalked. At the time, there were four people, including myself in the weight room. After I was there for a few moments I noticed there was an unusual amount weight banging. It was louder and more repetitive that it normally is.

I've been exercising and going to health clubs for about 15 years and I immediately recognized that these few people were repeatedly going out of their way to make as much noise as possible. I sat down on a bench and started doing dumbbell presses. Two students walked in, stood on either side of the bench I was sitting on and began to do lateral arm raises, *blatantly invading my personal space*. There was no special equipment that they were using in that area. I noticed one had a shirt on that said something like, "The MOST important part of playing rugby is support."

Anyone who can't wait a few minutes for you to finish and needs to stick their arms in your face is probably doing it on purpose. In addition, the exercise they were doing could have been done anywhere else a gym that was almost empty. It was a dead giveaway. Another student arrived and began lingering in an area to my left. He stared at me for a few seconds and started smirking.

Once he had my attention he turned his back to me so I could see his shirt which read, "CHASE." In smaller letters underneath was the word, "Manhattan." These attacks using symbolism are common reports among people who are targeted. The kid with the CHASE shirt left shortly after, and his total stay was about ten minutes. That episode is an example of noise, crowding, and symbolic attacks.

Noise Campaigns

All targets of the Hidden Evil are familiar with noise campaigns. Basically, targets report experiencing a steady stream of noise consisting of a rotation of various types of disturbances around their homes. This includes door slamming, yelling, car alarms, horns, tires screeching, loud music, engines revving, and frequent construction projects which include an assortment of noise from heavy machinery and tools. Not surprisingly, in his books *Future War* and *Winning the War*, Dr. Alexander described using noise as a weapon. In *Future War* he wrote, "Audible sound, in frequencies from 20 to 20,000 hertz, can be applied to influence behavior, as most people are sensitive to very loud noises."(*2)

For instance, if you're targeted, there will be doors slamming for a few moments, several minutes of silence will ensue, then a motorcycle will drive by and rev its engine outside of your house. After that you'll have someone mowing the lawn. Then there will be alarms going off. After that you'll hear a loud crowd of people walk by your house. Then you may then hear fire or ambulance sirens. Then there will be hammering and sawing sounds from the ongoing construction projects. Then you'll hear car alarms being turned on/off in succession multiple times from multiple vehicles around your house. There will generally not be a solid five minutes of uninterrupted peace.

Moret spoke of the use of "noise campaigns" which included the frequent use of garbage disposals in surrounding apartments, vehicles with "blaring horns" and the sirens of fire trucks. She also spoke of habitual door slamming, proclaiming, "they'll slam doors all day and all night long." "Door slamming," agreed McKinney, "is also a popular pastime, particularly in apartment settings." Those targeted may experience a "chain reaction" of door slamming by apartments or houses surrounding theirs. Other types of disturbances include frequent noise from lawn mowers, snow blowers, vehicles with loud exhaust systems, and ambulance sirens, etc. Noisy construction projects will usually be ongoing.

"Blaring horns, sirens, [and] garbage disposal[s]" which are "run concurrently in apartment settings, for excessively prolonged periods of time," are common reports wrote McKinney. She also noted that targets experience frequent "amplified transmissions" of "general racket" which is "used on a recurrent basis under circumstances intended to persuade the individual that he or she is under surveillance." She added, "In all of these cases, the individuals' neighbors apparently pretend to be oblivious and/or indifferent to these sudden, continuous explosions of noise." They're oblivious to the noise because they're complicit in the program.

Also, the people above a targeted individual may loudly pace as they mimic his or her movements from above. For instance, if you're targeted, as you walk into your bathroom, you hear loud stomping from above which follows you into the bathroom. McKinney affirmed, "A number of individuals report that occupants of upstairs and downstairs apartments appear to follow them from room to room, tapping on the floor or engaging in other activities which appear intended to advertise an ongoing surveillance." They'll "pace in their apartment," added Moret, and seemingly to the target underneath them, "it sounds like they're being tracked ... from the apartment above."

If you're targeted, helicopters and small propeller planes may frequently pass over your home. Some of these aircrafts may be low enough to vibrate the room you're in. According to Cheryl Welsh's March, 2003 article entitled, *List of Mind Control Symptoms*, the use of helicopters is a common tactic reported by targets. Apparently these overhead assaults are used for sleep disruption and to undermine the morale of targeted people.

Welsh is one of only six Non-lethal Weapons experts in the world recognized by the United Nations. This noise may follow you to destinations you frequently visit. Foot and vehicular traffic may be re-routed through your street, causing an unnatural amount of vehicles driving by your house (even if you're on a side or dead-end street). There also appears to be a fleet of vehicles equipped with loud mufflers that groups use specifically for creating noise in the targeted area.

Often, noise will be synchronized with some type of activity. For instance, a target hears an ambulance, fire sirens, alarms, or beeps as they leave. Or as they leave, a small aircraft passes overhead, then as they approach their car, they hear alarms or doors slamming. Noises made as they leave also appear to serve as a form of communication to alert group members in the area to begin pursuit. In addition to the standard noise package, whatever noise would naturally occur in the area you live will also be increased in frequency. For instance, if a target lives in a rural area then noisy chainsaws will be quite regular. And in the winter there may be frequent loud snowmobiles that pass near the area.

In this manner, noise is used as a weapon to inflict pain. Chronic exposure to even low-level noise is considered a health hazard that has been known to produce adverse physiological and psychological health effects. Prolonged exposure to noise can produce high blood pressure, a rise in cholesterol, damage to the circulatory, cardiovascular, gastrointestinal, and musculoskeletal systems. If you're targeted and pregnant or plan on having a baby, you may want to think twice. Noise has been known to cause hearing loss and growth disturbances in a fetus and can lead to birth defects.

Prolonged stress in general has been known to cause miscarriages, and some who have been stalked and harassed over a period of time have attributed their miscarriages to ongoing harassment. Just as the fabricated cut-offs are likened to getting *slapped*, these noises are the equivalent of being repeatedly *shocked*. The creators of this program are obviously aware that these noise campaigns are *acts of violence*.

Synchronization

Groups will try to synchronize their tactics with things that targets do such as entering or leaving their homes. The target's movements will be synchronized with vehicles or people coming or going, or other movements. These synchronization tactics are often done several times, perhaps three or more. For instance, when the target arrives, two cars will drive down their street, while their next-door neighbor leaves.

If a target has a habit of going to the bathroom at 2 AM and looking out his bathroom window at a specific area, then they will eventually arrange it so that there is some activity such as a light being switched on or someone leaving or arriving in that area. This is an example of the level of detail that they will resort to when tailoring a harassment protocol to the profile of a target.

Synchronization is also used with sensitivity tactics and noise. For instance, if you're targeted, you'll hear ambulance or fire truck sirens the moment you enter or leave your house. Or car alarms, hammering or some other type of noise will occur as you look out the window. As you leave your home, you may notice people walking by, wearing a color or holding an object that you've been sensitized to. If you pull into a parking lot, there will usually be cars arriving or leaving. Synchronization is also used with blocking/crowding. For instance, as you approach a store's entrance, people will be either walking in or leaving. They will often be wearing a color or holding an object that you have been sensitized to.

A variation of this is *echoing* and *mirroring*. Like other forms of non-verbal communication, *echoing* and *mirroring* are very powerful, (but usually unconscious) ways of interrelating. This form of information exchange has been outlined in the book, *Secrets of Sexual Body Language: Understanding Non-Verbal Communication*, by Martin Lloyed-Elliot. Most of us speak this language without even knowing it to attract others.

However, this form of communication has evidently been incorporated into this program as a tool to annoy targets. With *echoing* for instance, if a target is in a restaurant, or on public transportation, and they drink then an informant next to them will drink. If they pull out a cell phone, then the informant will produce a cell pone or similar device. When the target puts it away, so do they. In essence, they will mimic the movements of a target.

With *mirroring* for example, targets may frequently have a person or group of people walking parallel to them, in the same direction on the other side of a street, or any public area. Also, there may be someone directly in front of the target, and approaching the same intersection that they're approaching. The informants adjust their timing so they approach at the same time, and cross the street just as the target does, at which point crowding may occur. The informant(s) may also be wearing red, holding cell a phone or another device, which the target has been sensitized to. Dr. Alexander and Janet Morris also spoke of the use of *mirroring* to influence people, in their book, *The Warrior's Edge*.

Harassment Skits (Street Theatre)

Commonly used routes are used as avenues for various types of *Street Theatre* that unfold as targets pass through. *Street Theatre* is used during group stalking and contains verbal and non-verbal harassment, threats, insults, intimidation, and violence, conveyed overtly, or covertly using themes, symbolism, or other medium. They may be carried out after a target has been *sensitized* using NLP. This harassment can be relayed with metaphors, verbal remarks, and symbolism using clothing or other items. Targets may be stalked by a network of people who frequently stomp their feet, clear their throat or cough (even in the middle of summer) to indicate that they're part of the program.

Stalkers may be speaking on their cell phone and loudly accent the insult as they walk by and stare at targeted people. Blatant, hostile or smirking stares are also common. Sometimes they may not be on the phone and may just overtly insult you as they walk by. As an example, I have people walk by me and say things like *fuck you, fuck-up, fuck-head, you're disgusting*, etc. I've also witnessed them spit in my direction as I walk by them and have seen this spitting tactic used by teenagers who would spit on the lawn while I was on the front porch as they passed by my former residence.

Some words carry with them a particular *weight* to potentially *trigger* an effect on a person's emotional state. For instance, if you're at a restaurant you may have a couple sit down at a table next to you, and mention some events that are taking place in your personal life, in order to get your attention. After that they may make repeated references words such as paranoid, *crazy, scared, panic, insane, freak, sad, depressed,* etc. The idea is that the words will invoke the emotion. Not only is this possible, but it works best when you don't know it's being done to you.

This tactic appears to be a variation of a *pacing technique* used in hypnosis. It involves first linking something to the subject's ongoing experience (or personal life) and then using certain words intended to evoke a particular response. Morris and Alexander refer to these words as "verbal anchors" and state, "Verbal anchors can also be effective tools... Words such as *freeze* or *duck* transmit complete messages. ... Verbal anchors can elicit both positive and negative states."

In addition these words can be emphasized by changing their tone and volume when they're spoken, so that some words are slightly louder and spoken longer than the others. This can completely change the meaning of a message. This is called *metacommunication. Metamessages* are suggestions hidden in a statement, perhaps a compliment. The tone, volume, and rhythm of specific words in the sentence are changed so that the actual message is different than what the spoken words are.

"It's hard to defend against the anger and disapproval expressed in negative metamessages," stated McKay, Davis, and Fanning, in their book, *Messages: The Communication Skills Book.* "The attack is often so subtle that you aren't aware of exactly how you've been hurt..." The basic function of metamessages," they say, "is to say something covertly that you're afraid to say directly. Since metamessage attack is covert, there is little chance of overt retaliation."

These *metamessages* are quite similar to *indirect imbedded messages* used in hypnosis. In the book, *Patterns of Hypnotic Techniques of Milton H. Erickson, M.D.*, Richard Bandler and John Grinder provide examples of a master hypnotist named Erickson. Describing this type of covert communication, they write, "Imbedded commands serve the purpose of making suggestions to the client indirectly and, thereby, making it difficult to resist in any way... These are most effective when they are also marked analogically by emphasizing the command and by looking intently at the listener, if their eyes are open."

It appears a variation of this message can be conveyed with multiple people blitzing a person with cut-offs while talking on their cell phones, looking directly at the person and emphasizing (metacommunication) *weighted* words and/or information about the person's personal life that was obtained from surveillance. Groups will also use these phony conversations to harass targets in stores, restaurants, public places, elevators, etc. McKinney wrote of recurrent negative comments by strangers that are apparently intended to evoke feelings of "paranoia." She also wrote in her report that the stalkers operate in large groups over a long period of time—seemingly with the support of the government.

Bandler and Grinder describe a positive use of this tactic when quoting Erickson addressing an anxious patient, with "My friends tell me to *feel comfortable* and to *loosen up* when we are out on the town." Note that the italicized words are emphasized when spoken. This tactic can be used even if the intended subject is not being addressed directly, hence the name "Indirect Imbedded Command."

Again, something about the target's personal life or real-time ongoing experience is mentioned near him so he will identify himself as the subject, then *weighted* words conveying insults are used. These covert public mocking tactics using trigger words which are intended to evoke negative emotions, are based on NLP/hypnosis, and seem to be the result of an in-depth study of hypnosis by the CIA which occurred during the MKULTRA period.

"When we hear something, even from another conversation, we can't help but make images and sounds of it in our head," announce Adreas and Faulkner. "The professional persuader knows this intuitively. In NLP we know it explicitly." They affirm, "You cannot NOT communicate." The use of NLP/hypnosis by these citizen stalkers as directed by their handlers is obviously intended to drive people *insane*.

Victor Santoro explains how these tactics can be used in a destructive format, as a way to "foster your target's paranoia" and "suggest that he's being mocked..." He explains, "If you can get several accomplices to do this ... he'll get the feeling that mockery and ridicule of him are widespread." Santoro's book seems to have some accurate information pertaining to this program, which makes me wonder if at some point he was employed as an intelligence agent. At any rate, the Hidden Evil was obviously designed by experts in the behavioral sciences.

After a target has been *sensitized*, they can even be harassed, insulted, and threatened with symbolism using articles of clothing, newspapers, and other items. Just as body communication (Social Kinesics) is used to harass a target, so too is symbolic communication. Symbolism can be used with articles of clothing, especially during a wave attack where a target is blitzed by a crowd of informants who smile, laugh, stare at them, and cut them off as they pass. Some of the messages I've noticed on apparel include: *Stop and Die*, *Big Bro*, *Watch Your Back,* as well as Satanic, Freemasonry, and Wicca symbols. An attention-getting tactic may be used. It can be a cut-off, blatant stare, loud noise, or other sensitivity tactics. Remember how the perpetrators at the YMCA both used an attention-getting tactic? One man gave me a smirking stare and the other two swarmed me.

In essence, they will take an event that is unfolding in a target's personal life and make references to it using any number of methods. Interestingly, an article in the *San Francisco Bay Guardian*, on October 22, 2004, entitled, *The New COINTELPRO*, stated similar that tactics were being used on protesters at anti-corporate rallies. The theme may pertain to something in your personal life. They may play on this theme for a single day, a series of days, or weeks. In addition, they may choose a theme and try to project it into your life. Personally, I doubt that most of the citizens who are used to flash these messages are aware of their meaning.

Setups and Confrontations

Informants have apparently been used to frame targets for crimes or to participate in staged events which foster character assassination. A person may not be aware of some of these setups. Some staged events may include people provoking targets into confrontations. McKinney wrote about "Recurrent confrontations by unusually hostile strangers." "Seemingly homeless people who are well-dressed and very clean will get into confrontations with targets ... and create big public scenes," warned Moret.

McKinney described an episode involving a male target who spoke with a woman who had often paced in the apartment above. After he spoke with her about this pacing, which occurred even during her absence, she "immediately complained to the building manager that he was stalking her," wrote McKinney. "She conveniently forgot to inform the building manager that she had assiduously 'courted' this individual for several months, without success; and that she had been stealing his newspapers on a regular basis."

Presumably, these setups and staged confrontations are done to get targeted people into the penal or mental health system. At the very least they can aid with recruitment. Also, photos, or video footage taken of an enraged target after they have been antagonized (setup), may also assist with a smear campaign when leveraging support of the community to facilitate the removal of the target. Multiple complaints filed independently which entail the same report may also achieve this.

Reportedly, local police participate in stalking and have framed people for crimes. Referring to one target's repeated encounters with a police officer, McKinney wrote, "One of his recent acts was to 'frame' her with a drug possession charge. After pulling her off the road (a frequent pastime) and subjecting her to an illegal search (done, twice, so far), he conveniently managed to find a glassine packet of cocaine eight feet away, in front of his squat car."

Similarly, during the former Cointelpro, there were routine setups, and fabricating of evidence used to destroy a person's character or have them imprisoned. But this obviously hasn't stopped. According to the December 7th 2003 issue of the *Sunday Herald*, the FBI is on record using multiple informants to frame innocent people for crimes by having informants commit perjury. This can be done legally in the interest of "national security." The article sited a 141-page report by the House Committee on Government Reform, which stated that the FBI rigged evidence, and used informants to provide "false testimony," so that innocent people would be imprisoned. Oddly enough, some of these informants were *murderers*!

"Well over 20 murders were committed by FBI informants," the report revealed, and to protect them federal law enforcement "actively worked to prevent homicide cases [from] being resolved." The committee also concluded that "officials in FBI Director J Edgar Hoover's office were well aware that federal informants were committing murders." The article continued quoting Judge Nancy Gertner saying, "It is hard to conceive of accusations that shake the legal system closer to its foundation that would do more to challenge this nation's most basic assumptions of honesty, fairness, and trust in the administration of justice." She added, "All in the name of 'national security.'"

In his book, *The Franklin Cover-up*, former Senator John Decamp wrote, "In case after notorious case entirely unrelated to Franklin, Justice Department [DOJ] personnel appear as liars, perverts, frame-up artists, and even—assassins." Apparently Senator DeCamp experienced significant turbulence while investigating a child sex ring which included prominent individuals in Washington D.C. According to DeCamp, the elite used the FBI to orchestrate the cover-up.

Stores and Restaurants

Store and restaurant staff work in tandem with citizen informants to harass targets. Wait-staff in restaurants will be rude or give targets bad service. Or they may appear to be friendly as they stick their elbows in a targeted person's face, repeatedly kick their chair, or crowd them while taking an order or delivering food. They will also mess-up orders, and appear clumsy or incompetent by accidentally dropping food or silverware on targeted people.

If you're targeted and make a trip to the bathroom, you will be cutoff by wait-staff who appear from around corners or through doors. Plain-clothed citizens will be assisting them in the harassment as they crowd you while you walk to and from your table. They'll also follow you into the bathroom where they will crowd you or appear to be talking to someone on their cell phone while parroting information about your personal life.

In restaurants, these citizens will be seated around the target, wearing a color that he or she has been sensitized to. They may very well have conversations which contain specific information about the target's personal life. This is obviously a very empowering trip for some of them. This seems to be done, both to harass targets and to try to evoke a sense of helplessness in them by not being believed if they were to tell someone about the harassment.

Similar to restaurant staff, store clerks work with citizen informants in their store to crowd, block and cut targets off. They may engage in conversations with citizen informants at the checkout line or in the isles as a target walks by, which are obviously intended to be overheard, and contain information about a target's personal life. A variation of these tactics were used in East Germany, where the Stasi would co-opt a business owner to utter some words in the presence of a targeted customer, in order to let him know that they had been expecting him.

Stores and restaurants also participate in noise campaigns by banging merchandise on shelves as they put it away, throwing boxes on the floor as targets pass by, door slamming, slamming items on counters, and other types of commotion. If there are low ceilings, there may be banging directly above the target which will also follow them around the store. The *Noise Campaigns* and *Street Theatre* are apparently intended to create a cloaked hostile environment to persuade the target to leave, and let them know they're being watched.

Also, specific products that a target shops for may be sold out in the stores that they frequent. Apparently they are removed from the shelves prior to the target's arrival. In addition, if targets purchase products from vendors over the phone or online they will frequently encounter representatives who don't return calls, appear vague, or incompetent. Generally, targets will receive poor service. These things happen to all of us, but for targets it is recurrent.

Friends and Family

If you're targeted, friends and family may be recruited to break you down. They may participate in harassment skits intended to make you look incompetent, inconsistent, hypocritical, and to mock and demoralize you. The methods used to recruit friends and family will be covered in the *Informants* section.

New friends, may be used in an attempt to do the same. "A number of individuals in touch with us report a range of experiences with new 'friends,'" wrote McKinney. These 'friends' pose as confidants and abruptly end these 'friendships' under deliberately degrading and humiliating circumstances. She added, "When taken in the context of the ongoing surveillances and harassment, these exercises appear intended to heighten emotional trauma, perhaps to provoke an uncontrolled response and/or to enforce isolation."

If you're targeted and live with someone who has been co-opted, perhaps a friend or family member, they will probably be forced to carry out some of these harassment tactics. At home they may frequently slam doors, cabinets, draws, and make other types of noises. They may frequently crowd or block you. Areas that you commonly occupy, or spaces normally used, will be littered with objects (blocking), that you have been sensitized to.

Frequently used pathways will also be blocked. Traps may be set, such as items frequently falling out of freezers or cabinets as you open them. Phone conversations which occur while you're present will be loud and contain information about your personal life. Out in public, these friends or family members may be forced by their *handlers* to guide you through and participate in *Street Theatre*, where they will interact with other citizen informants to harass you. These skits will be intended to make you feel foolish, incompetent, or inconsistent.

Thefts and Break-ins

When break-ins occur there is usually no sign of forced entry, and they occur when the target is sleeping or elsewhere. This is the case even if there is a Hi-Tech security system in place. "In virtually all such cases [investigated], the burglars leave evidence of their visits, such as by relocating objects, or by committing petty and not-so-petty acts of vandalism," wrote McKinney.

This happens at the home as well as the workplace if the target is fortunate enough to be employed. Targets have reported clothing being ripped, milk, coffee and other items poured out, and small-scale damage. Items may be stolen and brought back at a later date, and they may move items and place them in a slightly different spot. These are essentially *Gaslighting* tactics.

Victor Santoro claims that these subtle memory games can be used to drive people insane. Small things such as pens, shoes, and silverware are taken or tampered with so people will seem delusional if they file a police report. "Burglarizing your home is very, very common," stated Moret, and "a lot of times they'll leave evidence of their visits, just little small changes in the target's home which is noticeable." Executive Order No. 12333 states that these break-ins can be done legally.

Interestingly enough, surreptitious break-ins, thefts, and sabotage were a staple part of the old *Cointelpro*. Other tactics Santoro advocates include filling a target's gas tank, or substituting items of his clothing with ones a size smaller. If you're unaware that you're targeted, as most people are, then these small changes may be attributed to a lapse in your memory. If these tactics are used in combination with others, then as Santoro illustrates, you may think you're going insane.

Gunderson also spoke of *Gaslighting* tactics, which he described as "psychological warfare." In his personal experience with what is presumably state-sponsored harassment, he has noticed bits of furniture and other items being moved around during surreptitious break-ins. He states they are, "so arrogant that they almost always leave subliminal messages of their ... entry." The logic is that a thief would never break into your home and leave a computer or stereo and just steal a toothbrush, or pour out a portion of your milk.

In his book, *Subliminal Mind Control*, John J. Williams wrote, "If you are being subjected to a secret and involuntary subliminal experience it is a malevolent, sinister and often dangerous attack against you that you have every right to vigorously oppose and thwart as if you are a POW or someone has broken into your home to rob you of what is rightfully yours." He warns, "In fact they have broken into your mind, and are trying to rob you of your free will. Even in cases where there is actually an intended positive outcome."

Sabotage, Vandalism and Staged Events

Targets may regularly experience acts of vandalism. Electronic equipment may frequently be failing. People have reported having electrical systems in their cars, brand new appliances, TVs, radios, PCs, and other electronic equipment suddenly die, or act strangely.

Targets may experience frequent computer trouble such as failing hardware. Ongoing computer problems which do not follow any logical pattern may also occur. Some of this can be attributed to the E-bomb which can destroy electrical equipment while not damaging humans.

Many times this vandalism is usually just below what you'd report to the police. For instance, a small but noticeable portion of your masonry has been chipped off, and a crack has appeared in one of your windows. This may be restricted to minor vandalism, just to let you know that you're under surveillance. However large-scale property damage is also done to homes and vehicles, such as slashed tires, broken windows, tampering with break lines, and destruction of electrical equipment.

Some targets have reported that their pets die suddenly of mysterious illness. For instance, they come home to find a perfectly healthy pet dead. Vandalism and pets being tortured may also be in retaliation for a target taking action to expose them.

"Automobiles are one of the biggest targets," said Moret, adding "they slash tires, they smash the windows, drain the oil out of the cars, fool with the electronic components and batteries." She continued, "A lot of people have complained about breaks suddenly failing, clutches failing." McKinney agreed writing, "Vehicles invite peculiarly ferocious attacks in these harassment campaigns—slashed tires, smashed windows, oil drainage, oil contamination, destruction of electronic components and batteries ... grounded fuel gages" and "suddenly failed brakes and clutches," occur. Recurrent auto thefts have also been reported.

McKinney explained that some of those she was in contact with had experienced staged accidents. One involved the breaks failing on a tractor which resulted in the death of the driver. And another episode included a person barely escaping being run off the road by an off-duty bus. "Two others," added McKinney "narrowly avoided what appeared to be deliberately attempted collisions by drivers who quickly sped away from the scene."

Another episode involved a targeted person who avoided three attempts in four days at being run off the road. Still another survived being run off the road twice within week, which resulted in a totaling of her two vehicles. "They also conduct staged accidents where people may be forced into accidents in their cars," echoed Moret.

So, according to this information, multiple seemingly independent vehicles can be used to facilitate an accident which can appear to have been caused by a targeted person. Add to this, multiple independent witnesses (informants) and you have a recipe for murder or a framing. In fact, "accidental" deaths have reportedly occurred. Once more, the Stasi would do similar things. "Car brake-leads have been cut, accidents and deaths reverse-engineered," noted Funder.

Traveling

When targets drive, they may be surrounded by vehicles driven by citizen informants who frequently cut them off and tailgate them. Targets will experience more tailgating and cut-offs than what would naturally occur. Vehicles will frequently pull out past stops signs as the TI approaches an intersection. This is apparently done to startle them.

While traveling on the highway there will be a rotation of vehicles that surround targets. These vehicles may be marked with an identifying feature, or a color that they've been sensitized to. This seems to function as both an NLP tactic to *trigger* negative emotions which have been *anchored* to the feature, as well as a method to unite the group and promote *group cohesiveness*. In other words, it is similar to a *uniform*, or a squad car.

There will also be a rotation of people tailgating targets. These vehicles may cut them off after they've been tailing them for a while. Apparently this is done to imply that they weren't going fast enough. When done in a recurrent manner by multiple vehicles, this antagonizing behavior may also serve to encourage the target to accelerate, which may result in them getting pulled over. This may be especially effective if they're not aware that they're targeted.

City and state vehicles which may stalk targets in public, include: off-duty metro buses, school buses, local restaurant delivery vehicles, city and construction vehicles, fire trucks and ambulances, police cars, postal and UPS vehicles, taxis, and 18 wheelers. Helicopters and small planes have also reportedly been used to stalk people. As will be revealed in the *Informants* section, many of these are vehicles belonging to organizations which have been selected for recruitment into the citizen informant programs.

Convoys of vehicles which are all spaced the same distance and traveling at the same speed will participate in stalking. Within this convoy, there may be several vehicles exhibiting traits that a target has been *sensitized* to with NLP. These traits are thereby symbolic of a *uniform*. For instance, most or all of them will be a shade of red. Or most of them will have their lights on, even during the daytime. Or they may have the same bumper stickers. During the evening targets may experience frequent *Brighting* while driving or walking.

Brighting

Vehicular *Brighting* is an attack. It seems to be strategically used with corners. For instance, if you're targeted and you're walking down a quiet street late at night, you'll see vehicles repeatedly turn onto or off of connecting streets. Their timing is such that they continually turn the corners while you are at it. This effectively blinds you. This happens frequently at corners. Or, as you walk down a side street, there is one or more cars parked on the side, facing your direction with the lights on. They will also wait for you to approach them and then pull out into the street, thereby *Brighting* you as you walk by.

These *Brighting* tactics are maximized when a vehicle has one headlight grossly misaligned. The misalignment produces a Hi-Beaming effect. Like other tactics, frequency and duration play an important role here. Lights being shined into a target's windows have also been reported. "Among the most common non-lethal weapons are bright lights," wrote Dr. Alexander, in his book, *Winning the War*. He says, "The intent is to reduce risks of injury ... by temporarily hindering the eyesight of the targeted personnel." He also wrote of the use of bright lights on targeted people in his book, *Future War*.

Blacklisting NO

Job opportunities will be trashed. Most targeted people are unemployed. Many would be in the street if it weren't for parents, siblings, or friends. Targets who are employed are usually Mobbed. "Progressive financial impoverishment, [is] brought on by termination of the individual's employment, and compounded by expenses associated with the harassment," wrote McKinney.

During her investigation she concluded, "The majority of those now in contact with the Project—educated, white-collar professionals—have lost their jobs. Termination of employment in many of these cases involved prefatory harassment by the employer and co-workers [Mobbing], which coincided with the other overt forms of harassment..."

According to an article in *The Christian Science Monitor* entitled, *Blacklisted by the Bank*, *Blacklisting* was originally used as a foreign-policy tool. At some point federal organizations obviously decided to start using it on individuals as well. Slander may play an important role in Blacklisting. Also there are Blacklists and software packages used to check these lists for suspected "terrorists," "threats to national security," etc. Apparently this is a flourishing business and some organizations are required by law to use it.

An August 2004 article entitled, *The Surveillance-Industrial Complex*, by the ACLU, stated "An entire industry has sprung up to produce software that makes it easier for companies to enforce the government's blacklists and other mandates. An example is 'Homeland Tracker,' produced by a subsidiary of the giant database company Choicepoint to 'help any business comply with OFAC and USA PATRIOT Act regulations.'" The article notes that the software has a feature called "accept and deny lists ... otherwise known as blacklists."

Federal law enforcement has sent these watch lists to corporations. They contained the names of individuals that were not under official investigation or wanted, but that the agency just had an interest in. We already know that some of these agencies are targeting and harassing citizens. Unbeknownst to many, these agencies are apparently deliberately destroying careers on a massive scale. *Wired NewsWire* spoke of these Blacklists on August 9, 2004, in an article entitled, *Big Business Becoming Big Brother*. The article concluded, "There is no way to determine how many job applicants might have been denied work because their names appeared on the list."

A form of Blacklisting has also been used on Russian dissidents by the state. In his book, *The Persecutor*, Sergei Kourdakov, a member of the plain-clothed People's Brigade declared, once a person was targeted, they were "treated like a leper and could only get the worst jobs."(*3) To accomplish this Blacklisting, the USSR used something called a *workbook*. The workbook was used to advance or destroy careers. If you were cooperative with the state, your career could be very successful. If not, you would live in poverty.

In East Germany dissidents were Blacklisted as well, and their careers were destroyed. After the wall came down they were allowed to inspect their files, and learned why they were unable to get jobs or enter into universities. Dr. Munzert, speaks of the similarity between what is now happening and what the occurred in Germany; he says the goal is "to bring their finances down," by making them "loose their job." Your "financial state is destroyed," added Moret.

If you've previously been Mobbed out of employment and are Blacklisted, the Mobbing may continue during interviews. Presumably, this is used to further traumatize you. For instance, if you're targeted, they may ask you a series of questions for which you have to answer "no" to, such as asking you for credentials that you did not list on your resume. They may ask you if you possess skill-sets which were not listed on the AD they had placed. These Mobbing tactics are apparently designed to humiliate you. These things happen to everyone ... once in a while. There is basically a "safe zone" that these people can harass you from with little fear of being caught.

Interview questions may be asked with accusing, or mocking tones. Questions may also be asked using certain key words, regarding the environment you just left, which only you and your former co-workers (who Mobbed you) would know. These questions are obviously designed to *trigger* negative emotions which have been linked to a previously painful work environment, where you experienced Mobbing. They may ask these questions over and over using a slightly different syntax.

If you're targeted, during an interview you may have potential employers demanding credentials that they did not list as a requirement in their AD. During phone interviews you may have employers telling you that you responded to an AD other than the one that you actually responded to. They may also tell you that they did not place the AD, or that they placed a slightly different one. You may be told that your updated resume says you're currently working at a place that you never worked at, or a place you worked at a decade ago.

The harassment is masked in layers of accommodation to help strengthen the illusion that you're not working because you don't choose to. For instance, you may have recruiters who have harassed you in the past leave messages on your answering machine or with relatives/friends, indicating that there are opportunities available to you. It will be made to look as if you're unemployed due to your choice. The general public may not realize that Blacklisting exists because the institutions that use it operate most major news outlets. Reporters and historians are also Blacklisted for reporting corruption or historical facts that the government-corporate complex does not want the public to be aware of.

Publishers Weekly, gives an editorial review on Amazon.com of a book entitled, *Into the Buzzsaw*, authored by Kristina Borjesson, in which they state, "The buzzsaw is what can rip through you when you try to investigate or expose anything this country's large institutions be they corporate or government want kept under wraps." This "buzzsaw" is what others have called the *Invisible Government*. The review continues, "Reporters who tattle risk losing their jobs and being blacklisted ... [by] concerted corporate and/or government efforts to kill their controversial stories and their careers."

Communications Interference

Targets experience tampering or lost mail. The target's "mail is intercepted," said Moret. There's theft and tampering." During the former version of Cointelpro, they would blatantly watch homes, follow cars, and open mail. The idea was to get as much info as possible to aid in the harassment and intimidation of their targets.

The Stasi too, "inspected all mail in secret rooms above post office," described Funder, "and intercepted, daily, tens of thousands of phone calls." Dr. Munzert noted that one objective of those running this program is to "disrupt the communication systems of the victims." McKinney and Moret state that targets usually receive lots of harassing telephone calls. These pranks may be linked to an event unfolding in the target's personal life. The phone has historically been an excellent harassment tool.

Online

Targets may also be stalked online. Search engine results may be filtered (blocked) and replaced with information in a target's personal life. Websites that a target visits may be spoofed and contain information which is apparently intended to let the target know they're being stalked in the internet as well. This appears to be simply an online version of the public stalking. If you have a website devoted to exposing this program, you may have people emailing you to flame you, or claiming that they are victims and asking for support while making references to subjects unfolding in your personal life. This tactic may be used to foster a sense of hopelessness and isolation, (i.e., you're the only one).

If you're targeted, you may receive "unsolicited" email that parallels a current event in your life. Or you may receive covert insults and threats. If you join a support group, you may also receive harassment via threads posted on message boards. Like other mediums of harassment, the topics of these threads may be about events that are unfolding in your personal life, as well as threats or insults covertly directed at you.

Your email may be blocked or filtered and your web activity may be monitored. Carnivore, DCS1000, or similar tracking software can accomplish some of this.(*4) If you try to sell things online, participate in online discussions groups, or respond to employment Ads, you will probably be interrupted, trashed, or harassed.

Some people have reported that their inbox is clogged with junk messages even if they have filters on. These unsolicited emails may contain cryptic borderline-subliminal attacks. The words will be garbled but contain letters that convey a message when arranged in the correct order. Or they may contain a *sounds like* message. Your mind will automatically piece these letters together so they make sense. This is done on the fly, without you being aware of it.

These messages may contain junk in the body and the TI may not be able to reply to it because it came from a bogus address. These cryptic attacks are mentioned in the book, *Subliminal Mind Control*, by John J. Williams. If you're targeted, the message will probably pertain to something in your life situation. Any weakness you have can be exploited. For instance, if you're going bald and have been looking for treatment, you could receive the following attack:

To: John Doe
Subject: Euro Paulding Tummy
Body: asdfl;jkasdflkkjasdf;lkjasdfljkasdfl;jkasdfl

TI TV

The harassment of some targeted people is apparently broadcasted around the country via closed circuit TV. In the workplace this can easily be done by surveillance systems, which many businesses now have in place. A report based on a survey conducted on over 500 companies appeared in the May 19, 2005 issue of *Business Wire* in an articled entitled, *Statistics Show Rise in Surveillance of Workers*. It stated, "More than half of the companies surveyed use video monitoring to counter theft, violence and sabotage," and that the number of businesses conducting surveillance is *increasing*.

McKinney has witnessed surveillance films of people being *Gang Stalked* in the office. She declared that it might be used for the purpose of "creating a sense of unity, [and] for identifying TIs who are to be harassed on the street." In addition to facilitating *group cohesiveness* and identifying targeted people, McKinney noted that it seems to be a "major source of entertainment" for the individuals participating in this harassment. Similarly, more overt degenerates have been known to record their actions so they can "re-live" the experience.

"Covert workplace surveillance is big business," stated the *Denver Rocky Mountain News*, on April 30, 2001. "Secret monitoring of employees has grown dramatically," echoed *Knight Ridder* in their August 6, 1997 article entitled, *The Boss is Watching as Workplace Surveillance Grows*. According to these publications, the equipment is installed after hours and includes miniature cameras hidden in thermostats, light switches, and on walls and ceilings that are located in changing areas, locker rooms, break rooms, and even bathrooms. Apparently there have been some lawsuits concerning this.

Non-lethal Weapons (NLW)

My goal here is not to cover this topic in detail but to provide evidence that these weapons exist and are being used on the civilian population.(*5) They are also known as, *Directed Energy Weapons* (DEW), *Psychotronic Weapons, Less-lethal Weapons*, and *Electromagnetic Weapons*. Most targets of The Hidden Evil are hit with these weapons. They are reportedly located in bases of operation which surround TIs. Some may be satellite or tower-based. People also receive DEW attacks while traveling, such as on airplanes or in vehicles. Apparently briefcase-sized portable weapons are used.

Some of these Non-lethals are *Through-The-Wall* (TTW) in nature. Some states, such as Massachusetts, have passed a law banning the illegal use of directed energy weapons. However, the law states that federal and local law enforcement can use these weapons. These weapons or similar ones are currently being used on a massive scale against citizens of North America and other NATO nations. Dr Munzert refers to "Directed energy weapons" as the "high-tech arms of the century," and adds that they are "part of crimes (in Europe) that almost nobody knows [about] except the victims and the offenders."

On the topic of *Directed Energy Weapons*, Moret stated, "this technology was outsourced to the FBI, and that was for the purpose of putting it in all the police departments in the United States, where it now is." "Congress back in the early 90s, late 80s, took the position that anyone complaining about these systems were imagining things," declared McKinney. "By 1992," however, "they were off the drawing board and in fact being fielded and conveyed to law enforcement agencies." Although the surface intention for the creation of these weapons is to save lives, the evidence suggests they are being deliberately misused on a global scale as a traceless form of *slow torture*.

Some types of attacks that targets have reported can be explained with conventional technology. For instance, some types of unclassified DEWs can cause nausea, fatigue, headaches, liquefy bowels and a variety of other symptoms. This is not science fiction. This technology currently exists. Some of it is over thirty years old and has been tested on civilians for decades. Although some of the technology being used on targeted individuals is probably much more advanced than the weapons outlined below, I listed them because their existence is undisputed.

In an article entitled, *Wonder Weapons*, the *U.S. News and World Report* stated, on July 7, 1997 that, "So-called acoustic or sonic weapons ... can vibrate the insides of humans to stun them, nauseate them, or even 'liquefy their bowels and reduce them to quivering diarrheic messes,' according to a Pentagon briefing." Under the title, *Surrender or We'll Slime You*, the February 1995 issue of *Wired Digital Magazine*, described that, "Enemy soldiers might be confused by holographic projections," and "disoriented by low-frequency acoustic beams that provoke vomiting and diarrhea." The article stated that although these "weapons may seem like they've sprung from the pages of Marvel Comics," that "research into these and other non-lethal technologies has been underway for decades."

Moret contends that some of these weapons are both land and space based. And McKinney spoke of the use of the electrical grid throughout the country, the use of microwave towers, and devices affixed to poles that are connected to power lines. She adds, "These weapons systems are used by neighbors surrounding persons who have been singled out as targets of opportunity." Despite some mainstream reports portraying this technology as "emerging," it has been used on the civilian population for decades.

Microwave

Microwave DEWs will produce dizziness, burning, headaches, eye problems, damaged nervous system and internal organs, heart attacks, strokes, aneurysms, cancer, and an inability to concentrate. "Their capabilities, generally, are to inflict pain in a highly focused fashion, and to alter mental states," proclaimed McKinney. "In amplified forms ... these frequencies have the capacity to kill."

A common microwave oven can be turned into a DEW by modifying the appliance to operate with the door removed. It can then be positioned against a wall to attack someone on the other side. You can also build a weapon for a few hundred dollars that will employ a more focused attack.

"The worst thing" says Dr. Munzert, are "the microwaves because they enter the body ... and hit the heart, and the brain, and testes." He asserts that victims are attacked with these weapons, "night and day," and says that those carrying out this secret policy are "merciless as you cannot imagine." Moret stated that these weapons "cause deteriorating health, digestive problems, tremendous pain ... sleep depravation and disruption." "Sleep disruption/deprivation," agreed McKinney, "is achieved by overt and electronic harassment."

Similarly, the Stasi allegedly used *Directed Energy Weapons* on their enemies. Dr. Munzert explained, "We have information that this Secret Service (in German you call it Stasi) used microwave weapons to dissolute ... political opponents." Funder stated, "The Stasi File Authority began to investigate the possible use of radiation against dissidents. What it uncovered shocked a people used to bad news." According to Funder, although the File Authority claimed they found no one had been killed with these weapons, they did find that "it was used with reckless disregard for people's health." Unfortunately, it seems that the File Authority was probably wrong.

An article entitled, *Dissidents Say Stasi Gave Them Cancer*, appeared in the *BBC* on May 25, 1999. It stated, "When three of former East Germany's best-known dissidents died within a few months of each other, of similar rare forms of leukemia, suspicions were aroused among their friends that this was more than just a coincidence."

Speaking of the current epidemic, Dr. Munzert mentioned that a personal friend of his was tortured to death and given cancer. In her book, *The People's State*, Mary Fulbrook, declared, "The Stasi not only initiated well-attested murders ... of a number of individuals but also attempted more subtle methods of causing long-term ill-health and death from less easily identifiable causes, such as cancers caused by exposure to sustained high levels of radiation."

LIDA

The Russian LIDA machine (patent 3773049) is an old brain entrainment device used for the drugless sedation of mental patients. This device can be used as a weapon to drain a person of energy. The pulse rate can be adjusted so that it causes fatigue or excitability (sleeplessness). Although the LIDA signal does not travel a great distance, it can be used TTW in an apartment building. This device is about the size of a breadbox and works silently.

"The LIDA machine was made in the 1950's by the Soviets ... [and] would put rabbits into a stupor at a distance and make cats go into REM... The Soviets included a picture with the device that showed an entire auditorium full of people asleep with the LIDA on the podium."
-Dr. Eldon Byrd, U.S. Psychotronic Researcher

Voice-To-Skull (V2K)

V2K, also called *Microwave Hearing* was discovered during WWII when soldiers noticed buzzing sounds while standing in front of an energized radar antenna. Dr. Joseph Sharp demonstrated a V2K success in the mid 70s while working for the Walter Reed Army Institute of Research. If you have a tightly focused antenna, this can be transmitted through walls and over a distance. Voice or other sound can be transmitted over a distance, through walls, directly into the skull of a person. This was possible more than 30 years ago.

The March 1975 issue of *The American Psychologist*, which is the *Journal of the American Psychological Association*, declared that 'wireless' and 'receiverless' communications of SPEECH," had been achieved by Dr. Sharp. "Communication has in fact been demonstrated," they proclaimed. "The capability of communicating directly with a human being by "receiverless radio" has obvious potentialities both within and without the clinic."

McKinney wrote, "The Walter Reed Army Institute of Research (WRAIR) has participated in this research since Project Pandora. In 1973, WRAIR discovered that externally induced auditory input could be achieved by means of pulsed microwave audiograms, or analogs of spoken words' sounds. The effect on the receiving end is the (schizophrenic) sensation of 'hearing voices' which are not part of the recipients' own thought processes."

V2K is used to taunt and mock targeted people. Targets of The Hidden Evil have reported hearing doors slamming, voices, degradation of a religion, screaming, etc. Some have experienced mock executions with gunshots. Regarding this being a common tactic, McKinney stated, "the inducement of auditory input [V2K]," is typical. "A lot of victims" report "Microwave Hearing" which is a "scientific fact," announced Dr. Munzert. He describes these V2K attacks as a type of *torture*.

Alex Constantine was aware of these weapons and their intended use when he wrote in his book *Virtual Government*, "The intelligence agencies are capable of transmitting voices [V2K] and images directly to the cranium's sensory pathways. Anyone who falls into disfavor with this elite can be condemned to interminable physical and psychological torture. The victim is often murdered without a trace."

Silent Sound

Silent Sound (patent 5159703), developed by Dr. Oliver Lowery of Norcross, Georgia, also called *Clear Channel*, is an improvement over the time-slicing subliminal suggestions. It does not need to compete with other sound to influence a subject because it occupies a separate channel just beyond the human hearing range. Hence the name, "Clear Channel." This channel is beyond human hearing but not human perception. It is not normally in use but it is a direct conduit to the subconscious.

Silent Sound can be used by installing a hidden speaker near person's area of work or in their home. These suggestions can be played constantly and the person would not be consciously aware of them. Apparently this weapon was used on Iraqi soldiers during the first gulf war. Silent Sound can be transmitted through ordinary radio or television carrier frequencies. Only a speaker is necessary. But Silent Sound cannot only convey suggestions—*it can transmit cloned emotions*!

An articled by Judy Wall entitled, *Psy-Ops Weaponry Used In The Persian Gulf War*, which appeared in the October/November 1998 issue of *Nexus Magazine*, stated, "Subliminally, a much more powerful technology was at work: a sophisticated electronic system to 'speak' directly to the mind of the listener ... and artificially implant negative emotional states—feelings of fear, anxiety, despair and hopelessness." Shockingly, the article continued, "This subliminal system doesn't just tell a person to feel an emotion, *it makes them feel it*; it implants that emotion in their minds" by the transmission of "cloned emotional signatures, the result is overwhelming."

Silent Sound can also be piped through the V2K medium and transmitted over a distance, according to retired engineer and Targeted Individual Eleanor White. So this means that subliminal messages can be transmitted over a distance, through walls, directly into a person's skull. In essence, into their subconscious mind. "All schematics, however, have been classified by the US Government and we are not allowed to reveal the exact details," stated Edward Tilton, the President of Silent Sounds, Inc. He announced, "The system was used throughout Operation Desert Storm (Iraq) quite successfully."

DEW harassment may be synchronized with other types of harassment. For instance, if you're targeted, just as your computer crashes or the power goes out, you get V2K. Or as you hold a letter from the IRS, you receive a shock to your hand. This is apparently done for the purpose of mocking you. As if to convey the message, "ha ha, it's from us."

Some targets have reported being hit with other more exotic types of weapons that cannot be explained with current unclassified technology. These weapons have been known to penetrate all types of shielding.

The Cincinnati Post ran an article entitled, *U.S. Taps High-tech Arsenal/Pentagon Likely to Debut New Weapons in Iraq*, on March 30, 2003, which included an interview with Clark Murdock, a former Air Force Strategic Planner, who works for a Think Tank called the *Center for Strategic and International Studies* (CSIS). Murdock declared, "Once you're engaged and you have a capability that's almost ready, you'll try it." He continued, "All kinds of things have been invented, particularly in the (classified) world that will be used. If you use it and it works and no one knows, why talk about it?"

When I consider who runs this planet, what they've done, and what their obvious intentions are, it becomes clear to me that they intend to use these weapons to defuse threats to their control. For instance, if someone is a particular nuisance to the elite, they can arrange for them to experience frequent confusion. Unless this individual knew that they were targeted, they'd be completely unaware that this trouble was the result of an attack. On the other end of the spectrum, if the situation calls for the immediate neutralization of someone, they can be given cancer, a heart attack, or stroke. This type of murder is completely silent and invisible. Dr. Munzert refers to it as "the perfect crime."

Support Groups

My observation is that some of these support groups have been infiltrated (or created) by perpetrators posing as victims, used for discrediting, disinformation and disorganizing—*the triple D*. If you think that the people who oversee this worldwide program have not infiltrated these groups, or even deliberately created some as *catch-nets* in order to disrupt and minimize progress, you are probably mistaken.

The people who designed this system are highly intelligent and some of these support groups seem to be just another phase of the program. The East German Stasi would create political groups and foster an informant's rise to respectability. This influential informant would then impede the progress of the group and misdirect its members at events. The logic here appears to be that if these groups must exist, they would rather control them.

Some of these perpetrators seem to be very vocal and popular members of these support groups. It seems that this a *damage-control* mechanism put in place to corral people, manage them to some degree, and impede the groups' progress. These people may also help with misdirecting events, or generally keeping groups disorganized and ineffective, under the illusion that progress is being been made.

Guided by their handlers, multiple prominent informants may work in unison to jacket or otherwise discredit targets, who the organizers of this program believe are a hazard to its exposure. Jacketing was often used during the old *Cointelpro*, and involves the use of informants within an organization who portray effective targets as informants. We can only conclude that in the last twenty years they have perfected this tactic. If you are raising awareness, then discrediting attempts will be standard practice. It appears to be critical that they isolate you from group members who you may have a positive influence on.

Similar to the East German informants, they may be installed and built up to degrees of respectability by providing useful information to the group and engaging in other activities which genuinely damage this program. Although this may seem like a contradiction, this is the chess equivalent of sacrificing a bishop to take a queen. They know some targets would have eventually found the information anyway, so this trade-off is worth appearing genuine and gaining trust, which may be exploited at a later date.

It may also create fear and uncertainty within some targets, causing them to doubt their own judgment. Finally, this may further traumatize a target with feelings of hopelessness when they learn that a very well respected group member is harassing him/her. It's plain to me that these people exist to expose a portion of the truth, but to help destroy those who seek to expose all of it.

In my opinion, in order for a decent person to be harassing victims, especially when he or she has an idea what they're going through, they would have to have been blackmailed or tortured into becoming informants. If you choose to participate in one of these support groups, you may want to limit your exposure to certain people. However, although these groups have their share of perpetrators, not all of them are. So, if you're targeted you may still want to attend meetings and events as it will be a good opportunity to connect with other people.

Summary

When used in combination on a recurrent basis, these tactics are a painful and difficult to prove form of *torture*. Targeted people are battered from state to state and brought to financial ruin. Those who speak out may end up in mental institutions. Some are framed for crimes, or setup in a manner to aid with recruitment. Their homes are vandalized, pets tortured and killed. The mental, emotional and physical torture that targeted people experience will usually not stop until death. The creators of this system have established a virtually traceless protocol for *murder*.

Endnotes

* This Idea was taken from Gary Allen's book, *None Dare Call it Conspiracy*.

*1 Words such as "*crowding*" and "*space invasion*" have not been specifically mentioned in the material which I've found relating to the East German and Russian citizens watch networks. So this seems to be a new tactic, which is now used worldwide by this global stalking group. Although, the book, *Stasiland*, by Anna Funder, does hint that at least one target regularly experienced this type of crowding attack.

*2 He also wrote of the use of stink bombs to induce gagging or vomiting, and infesting the dwellings of targeted individuals with insects.

*3 Sergie Kourdakov defected to Canada and eventually ended up in Los Angeles, California. In addition to writing his book, he spoke in churches, on television, gave newspaper interviews, and spoke before government officials describing the practices of the Soviet secret police. He reported to his friends that he had received death threats. His plans to speak to the Russian youths by way of radio broadcasts were cut short by his sudden death on January 1, 1973, which was first ruled a suicide, then an accident.

*4 There may be a more advanced type of technology, which is not available to the public, that they're using to accomplish this online harassment. More on this will be covered in a future book or article.

*5 A more detailed description of this unclassified technology can be found at Eleanor White's website, www.raven1.net.

Chapter 18

Informants

An Ancient Phenomenon

The use of citizen informant networks dates at least as far back as the Roman Empire. *Delatores* (informants) were recruited from all classes of society, including knights, freedmen, slaves, wealthy families, philosophers, literary men, court officials, lawyers, etc. Similar to the TIPS program, it was an "all hands on deck" approach to empire security. Setups were routine, and informants sometimes received a portion of the land of those who they helped destroy.

More recently the fascist dictatorship of Portugal used the International and State Defense Police (PIDE) as the main instrument of political oppression. It consisted of secret police and a vast network of *Bufos* (plain-clothed citizen informants), who were apparently on every block. Money and a need for recognition (a pat on the back) motivated them. Second only to the Stasi in its thoroughness, the PIDE neutralized all opposition to the dictatorship.

Other countries have used massive citizen informant networks to destroy perceived opposition to dictatorial rule as well. In Czechoslovakia they served the Czechoslovak State Security (StB), and in Poland they worked for the Ministry of Public Security (MBP). The citizen informants of the State Protection Authority (AVH) ensured the survival of the Hungarian dictatorship. Targets were harassed, threatened, confined to mental institutions, tortured, blackmailed, and framed. Even their friends and family were co-opted to persecute them.

Probably the best recent example of citizens Gang Stalking people on behalf of the state is East Germany. In Germany the plain-clothed citizen informants were called *IMs* (inofizielle mitarbeiter), or "*unofficial collaborators*." Unofficial means that they unofficially worked for the Ministry of State Security (MfS), also called the Stasi. For her book, *Stasiland*, Funder interviewed former IMs, targets, Stasi Psychologists, and professors who trained IM recruiters in *Spezialdisziplin* (the art of recruiting informants). "The IMs," wrote Funder, "were 'inofizielle mitarbeiter' or unofficial collaborators [plain-clothed citizen informants]."

Funder continued, "In the GDR, there was one Stasi officer or informant for every sixty-three people. If part-time informers are included, some estimates have the ratio as high as one informer for every 6.5 citizens." In his book, *Stasi: The East German Secret Police*, John O. Koehler agreed that when you add in the estimated part-time IMs, "the result is nothing short of monstrous: one informer per 6.5 citizens."

In Russia, the People's Brigades were told, "We have growing problems in our country with enemies of the state. They operate internally, intending to undermine the authority of our government." Sergei Kourdakov tells how he was used to harass enemies of the state in his book, *The Persecutor*.(*) These specialized groups were referred to as the *Voluntary People's Brigade*. They were given a "License to harass," and charged with the "Maintenance of Civil Order."

When they sprung up all over Russia they were directed by plain clothed police on orders from Moscow. The citizens were told that they were part of a "special-action squad" and would be given tasks that the regular police couldn't or didn't have time to handle. They wore regular street cloths, and were convinced that they were ordinary citizens aroused into taking action against undesirables. Their leaders informed them that some of the people they'd be harassing were worse than murderers.

So in Russia and Germany, these informant groups were basically told that they were their country's first line of defense against threats to national security and criminals. But where are they getting the hordes of citizens who surround targets in public today? Well, they probably do pick people off the streets and use door-to-door recruitment. But due to the shear number of individuals now involved and the pervasiveness of this program, there must be a blanket recruitment process.

A Familiar Pattern

Targeted people have reported that everyone from homeless people to white-collar workers are participating in Gang Stalking. This also includes neighbors, friends, co-workers and even family of targeted people. Federal, state and local governments are reportedly complicit, such as local police, fire departments, EMT personnel, city workers, utility companies, taxi drivers, security guards, and stores and restaurants. According to Dr. Kilde, other participants include, "'Down and out' people, jobless, freed prisoners, mental outpatients, students and orphans." These people "are trained by this organization to harass, [and] follow ... innocent people, who for whatever reason have been put on the organization's hit list."

The April/May 1996 issue of *Nexus Magazine* revealed that "Tens of thousands of persons in each [Metropolitan] area," are now "working as spotters and neighborhood/business place spies (sometimes unwittingly)." There is a legal loophole that allows people to operate on behalf of the government, without knowing it. Executive Order 12333 states that organizations used by U.S. intelligence do not need to know that they serve U.S. intelligence objectives. These spotters, it continued, are charged with, "following and checking on subjects who have been identified for covert control by NSA personnel." This helps to explain the Mobbing and Gang Stalking accounts of targets both in the workplace and in public.

In order to *sell* the community on this program, its creators probably equipped it with convincing propaganda, which is delivered by professionals. Although there has been no official admission that the following federal resources function as recruitment programs for public harassment, they are similar to ones used in German and Russian dictatorships.

The Terrorism Information and Prevention System (TIPS), appears to be a bulked-up version of a community-policing program. Even though TIPS was officially rejected by congress, the American Civil Liberty Union contends that it and similar programs are being used aggressively across the nation. Other programs must have sprung up around the planet because the reported tactics of Gang Stalking in other countries are similar.

But who is the threat? According to some of these documents at www.citizencorps.gov, the targets are criminals and terrorists. However, the definition of a terrorist in section 802 of the US Patriot Act is frighteningly vague. Basically it defines terrorism as any action that endangers human life or that violates state or federal law.

An article entitled, *US Planning to Recruit One in 24 Americans as Citizen Spies*, which appeared in the *Sunday Morning Herald* on July 15, 2002, provided us with another clue. It stated, "The Terrorism Information and Prevention System, or TIPS, means the US will have a higher percentage of citizen informants than the former East Germany through the infamous Stasi secret police."

The articled revealed that the system is poised to recruit those whose work provides access to homes, businesses, and public transportation systems. Postal workers, utility employees, truck drivers, train conductors and others are to be recruited. On "state and local" levels these informants are to be directed by FEMA. This provides a motive for the reports of people being stalked by 18-wheelers, busses, city, postal, and utility vehicles.

It continued, "Informant reports will enter databases for future reference and/or action ... [which] will then be broadly available within the department, related agencies and local police forces. The targeted individual will remain unaware of the existence of the report and of its contents." This also helps to explain why inquiries by targeted individuals regarding the harassment have been met with denial by friends, relatives, neighbors, law enforcement, and colleagues.

In August, 2004, the ACLU, published a report entitled, *The Surveillance-Industrial Complex*, in which it contended that there is currently a vigorous citizen informant recruitment process. It stated, "Only under the most oppressive governments have informants ever become a widespread, central feature of life." Recognizing a familiar pattern, they charged, "The East German Stasi ... recruited from among the citizenry ... as many as one in every 50 citizens, to spy and report on their fellow citizens." They warn that a "massive" recruitment effort is underway.

The New American stated in their October 7, 2002, article called, *TIPping off Big Brother*, that the current effort to build a colossal network of informants is being done to, "enlist American citizens in surveillance activities that the state is either legally or physically unable to do." They say, a society, "where neighbors, co-workers, and passersby are all enlisted in a vast network of civilian informants—resembles conditions that existed in Stalin's reign of terror, and in all modern totalitarian states."

Continuing, they add, "Realizing that a cowed and brainwashed populace can carry out surveillance better than a million trained agents, Communist tyrants from East Germany to Cuba created revolutionary circles, youth groups, and other organizations specifically to enable the Party faithful in every walk of life to police everybody else."

On July 15, 2002, the *Washington Times* ran an article entitled, *Planned Volunteer-Informant Corps Elicits '1984' Fears*, which stated, "1 million informants" would be "initially" participating in a citizen informant program. And that the program would involve a combined effort between the DOJ, local police forces, as well as state and local agencies (businesses). "At local and state levels, the program will be coordinated by the Federal Emergency Management Agency," they described.

"Critics," they added, "say that having Americans act as "domestic informants" is reminiscent of the infamous Stasi" which targeted "dissidents and ordinary East German citizens..." This evidence helps to explain how stores and restaurants are participating in the harassment. In addition, it appears that these informant squads are coordinated by local and federal law enforcement.

According to www.citizencorps.gov, the directive of the informant program is to "harness the power of the American people by relying on their individual skills and interests to prepare local communities to effectively prevent and respond to the threats of terrorism, crime, or any kind of disaster." This may explain the reports of targeted people who allege that local, state and federal workers are using skills within their profession as part of the harassment.

For instance, city vehicles, postal vehicles, fire trucks, school buses, and taxis are reportedly stalking people. Construction projects encircle a targeted person's home and also spring up at frequently visited places. Utility companies interrupt service. Local businesses provide poor service, appear incompetent or clumsy and work with civilian informants to harass targeted people in their stores.

In East Germany the IMs included doctors, lawyers, journalists, sports-figures, writers, actors, high officials in religious organizations, pastors, waiters, hotel personnel, and other workers. "Schools, universities, and hospitals were infiltrated from top to bottom," wrote Koehler. In other words, what Koehler is describing is that the controlling faction of the organization was recruited into a policy set by the Ministry for State Security (MfS), which was then filtered down to its workers.

Interestingly, when I asked a manager at a retail establishment in Medford, MA if he had heard of Gang Stalking, he told me to "contact corporate headquarters." In all likelihood these programs originate from the organization's Corporate Headquarters that have adopted it as part of a federal or state policy. The pattern that is unfolding is that all major departments of the community are involved.

Other questionable programs include VIPS, Weed and Seed, Cat Eyes, and Talon. Part of the Weed And Seed program calls for a combined effort between local, state, and federal agencies, as well as community organizations, social services, private sector businesses and residents, to "weed out" undesirable individuals.

In a May 14, 2003 article entitled, *Building a Nation of Snoops*, the *Boston Globe* stated, "Watching America with Pride, not Prejudice," is " the Orwellian motto of the New Jersey-based Community Anti-Terrorism Training Institute, or CAT Eyes." CAT

Eyes is, "an antiterrorist citizen informant program being adopted by local police departments," spanning from the east to west coast.

The *Globe* described how this informant program was poised to recruit *100 million citizens*! The program will, "dwarf the citizen informer programs of the most repressive totalitarian states, making them appear amateurish by comparison," they wrote. "Even communist East Germany," they proclaimed, "was not as ambitious about citizen surveillance as CAT Eyes." According to the *Globe*, the goal is to recruit one out of every three citizens.

Catherine Epstein at the Department of History at Amherst College, contends that East Germany had the highest agent/informer to population rate in history. So if the U.S. has implemented this or a similar program, it has set a historical record. Former FBI Special Agent Ted L. Gunderson indicates that eventually about one in ten people will unofficially work for the state as an informant. In Germany the recruiters were given quotas to ensure a minimal amount of expansion. Due to the shear number of people now participating, I suspect quotas are being used.

<u>They've Recruited the Youth</u>

Children are now participating in the Gang Stalking of targeted people. Infants too, are dressed in colors that targets have been sensitized to. They are used as a billboard for symbolism, as they're paraded around by their parents who encircle targets in public. As demonstrated, after a target has been alerted that they're being watched, this color essentially acts as a uniform and a weapon. Later as the infants can walk and communicate they're used in other types of Staged Events (Street Theatre). Even at young ages, these children are able to understand how to perform these skits.

This program appears to be multigenerational, for both targeted people and the stalkers. Reportedly, entire families are now participating in patrols (Gang Stalking). Similarly, according to the book, *The File*, by Timothy Ash, sometimes entire families were recruited as IMs in East Germany. At this point we don't know exactly what these adults are told to encourage their children to participate. But by their behavior, we can logically conclude that the same factors which motivated them are used to get them to recruit their children.

Reason indicates that after a parent has been recruited, a message similar to, "there are bad people in the community and mommy and daddy need your help," is conveyed to the child. The child then receives some type of training and is used for patrols (Gang

Stalking). Family participation also seems to serve as a type of bonding, summed up by the adoption of attitudes such as, *let's work together to keep the community safe, we're the good guys, etc.* Consider that in East Germany, an astounding 6% of the IMs were children.

Due to the number of youths participating, there must be a broad recruitment program designed specifically for them as well. These programs are probably offered in high school and junior high school. In 2000 and 2001, *University Wire*, ran articles such as, *Boston U. Adopts Crime Watch Program*, and *Police Enlist Ohio State U. Students in War on Crime. The Atlanta Journal and Constitution* published articles in 2001 entitled, *School Police Enlist Aid of Students in New Crime Watch*, and *Student-led Crime Watch on Duty at High School. Students Keep the Peace* appeared in the October 2000 issue of the *Sarasota Herald Tribune*. Some of these policing groups are restricted to school grounds while others are not. Similarly, children were also recruited as IMs in East Germany while at school.

But the organization which appears to the central point of recruitment for youths, is the *Youth Crime Watch* (YCW), which is international in scope, with headquarters on five continents. It is partnered with the Citizen Corps' *National Neighborhood Watch* program, and works with adults to patrol the streets. Schools are used as recruitment grounds for students from grammar school up to the university level. Participants include all core components of the local government, such as law enforcement, school staff, businesses and restaurants, as well as adult citizens watch groups. After a period of training, they conduct youth patrols under the supervision of a nearby police officer or plain-clothed adult. Like their parent organization, they're fighting crime and terrorism. They have at least hundreds of thousands of members.

In addition to being used to carry out some of the standard tactics, the use of children may also be intended to disgust TIs.(**) This appears to be a triple attack. First, the act of someone using a child seems to be intended to cause revulsion. Secondly, the understanding that a TI would probably not be believed for even implying that a parent would involve their child in something like this, appears to be intended to amplify the revulsion. After all, nobody would ever, under any circumstances, use their children like this. And thirdly, the act itself. Most of these parents seem to exhibit no disgrace, which leads me to conclude that they believe it's a necessary honor.

If children are being used in this program, and if there is a faction of it which frames people for crimes, then is it possible that this faction would use children to frame targeted people? All that the parents and child would need to be told is that their country needs them to do something very important as a matter of national security. If they're sufficiently naive, they'll probably consider it an honor. Children of criminals, or those vulnerable to blackmail or bribery, can also be recruited to accomplish this.

A Nation of Stalkers

Most of the citizen informant programs previously mentioned are now organized under a National Neighborhood Watch program, known as, *USAonWatch*. "USAonWatch is the face of the National Neighborhood Watch Program," declared the Citizen Corps website. "The program," it announced, "is managed nationally by the National Sheriffs' Association [NSA] in partnership with the Bureau of Justice Assistance, Office of Justice Programs, [and] US Department of Justice." So far, this information suggests that the program is coordinated locally by the FBI and police.

The National Neighborhood Watch Program has existed since the late 1960s. This helps to explain reports of people being Gang Stalked by citizen patrol groups since the early 1980s. "For more than 30 years," proclaimed the National Sheriffs' Association, "The National Neighborhood Watch Program, an initiative of NSA, has been one of the most effective ways for citizens to become involved with law enforcement for the protection of our neighborhoods."

Neighborhood Watch is now partnered with Citizen Corps. And Citizen Corps is run by the Federal Emergency Management Agency (FEMA) under the Department of Homeland Security (DHS). Its website states, "Citizen Corps is coordinated nationally by the Department of Homeland Security," and "works closely with other federal entities, state and local governments." So in essence, USAonWatch is a Department of Homeland Security program. This means FEMA, DHS, the FBI, and local police are working directly with the state and local governments to coordinate these operations.

So if USAonWatch is the current public front, then that would mean that a portion of the Gang Stalking which targets speak of is done during these citizen patrols. It would also signify that this network, *with full complicity of all levels of government*, is trained and directed by local police, working with FEMA and the FBI. Again, we find another parallel to the Russian and East German dictatorships. The

National Sheriffs' Association exclaimed, "The Neighborhood Watch is Homeland Security at the most local level!" It was created to "empower citizens to become directly involved with Neighborhood Watch for the purpose of homeland security." Its targets, according to the NSA, are criminals and domestic terrorists.

What else would the East German IMs do besides inform? Markus Wolf, a former East German executive Stasi officer, and author of the book, *Man Without a Face*, indicates that targeted people were Gang Stalked by "informers who literally encircled their everyday movements." According to Wolf, the residences of some targeted people were "put under siege," and every family member, and every visitor, was kept under close observation. Even some East German state officials were, "surrounded by unofficial informers for the Stasi," proclaimed Mary Fulbrook, in her book, *The People's State*.

The book, *Stasiland*, reveals a similar citizen patrol scenario, where targeted people would be stalked by a network of IMs on a daily basis. This was done using a rotation of IMs. "In the morning when I went to work, there'd be someone close behind me," reported one target. "[And] if I went in to Alexanderplatz to do some shopping a man would come with me..." People would be followed by IMs from their doorsteps, into public transportation systems and then back home again. "They changed the personnel," she continued, "but there was always someone there. They wanted us to feel it." Despite what these IMs were told about threats to national security, this blatant Gang Stalking was done for intimidation.

In 1974 the former Deputy Director of the CIA, Ray S. Cline, drafted a report published by the Center for Strategic and International Studies (CSIS), at Georgetown University called, *Understanding the Solzhenitsyn Affair: Dissent and its Control in the USSR*. It stated, "The Committee for State Security" [KGB] ... functions as a secret political police force, which through nets of agents and informants, reaches literally into every crevice of society." The word, "literally" is noteworthy, because when it is used, they mean it. Just as with East Germany, you can bet that this means every major area of society was filled with informants. This would include workplaces, grammar schools, universities, businesses, restaurants, apartment buildings, etc. Literally, everywhere!

The report continued, "By watching every critical echelon in society, the informant networks also serve to insure that the administrative or daily controls are in fact working." The Soviet Union was run by a small group of wealthy elite. They used their private security force, the KGB, to ensure their control. Although the surface claim for the existence of the networks was for state security, it was in reality used to terrorize the population into submission. "These informants," it sustained "serve the fundamental purpose ... of inhibiting people from speaking freely to one another; from speaking out; from sharing and germinating thoughts."

Shockingly, according to an independent media report, there was an attempt to integrate Markus Wolf and KGB General Yevgeni of Russia, into federal law enforcement. An article entitled, *Ex-KGB and STASI Chiefs to Work Under Chertoff*, appeared in the *Foreign Press Foundation* on December 16, 2004. It explained how the U.S. would pay former Stasi officer Markus Wolf, and KGB General Yevgeni Primakov to help co-op citizens into participating in state-sponsored harassment.

The article was based on a U.S. Department of State Daily Press Briefing, that occurred on June 9, 1997, which stated, "We're talking about Markus Wolfe (sic), the former head of Stasi, right? ... He actively aided and abetted and fostered international and state-supported terrorism when he was an East German government official..."

So, when they sprung up in Germany, they were called "IMs" who unofficially worked for the "Ministry of State Security" to "protect the state" against dangerous elements. They were called the "People's Brigade" in Russia, and their duty was the "Maintenance of Civil Order." The current citizen squads are apparently fighting terrorism and crime. I'm not aware of the official names of groups which conduct these patrols in other countries; no doubt the participating countries have assigned them noble titles.

According to this information, those who have been recruited into the current citizen informant program include people who have access to homes, businesses, and public transportation systems. Others include postal workers, utility employees, truck drivers, train conductors, local police forces, as well as local and state agencies. And, if these youth crime-watch programs are being used for Gang Stalking—children. This helps to explain the reports of targeted people who attest that a variety of local and state vehicles are being used to stalk them. Restaurants and retail stores are also involved. Employees, clerks and/or managers of these establishments participate. If mainstream publications are alerting us to a "massive" recruitment process, then what they're doing behind the scenes is probably much more intense.

All of these volunteer Neighborhood Watch programs are sponsored by the *Department of Homeland Security* (DHS). Interestingly, in East Germany, the adults and children who Gang Stalked targets operated on behalf of the *Ministry for State Security* (MfS). In Russia, citizen informants worked for the *Komitet Gosudarstvennoy Bezopasnosti* (Committee for State Security), also known as the KGB. In Czechoslovakia it was the *Czechoslovak State Security* (StB), etc.

Because these harassment programs involve a combined, coordinated effort between federal, state and local governments, logic would suggest that the chief of police, mayor of the city, and governor of the state *must* be aware of it. The program involves the use of too many resources for them not to be. It *can't* operate without their knowledge. As previously described, the TIPS or USAonWatch programs would have accomplished this.

East Germany had a variety of different types of informants, which served different functions. So, in all probability, in addition to bulk recruitment programs such as TIPS, and Neighborhood Watch, there are individuals and groups of informants which are also recruited. In my estimation, these units and individuals work alongside the public front organization to harass people in public. If this is true, then this would further compartmentalize these operations, possibly to the point where all of these factions are unaware of the existence of one and other.

Consider the FBI's known use of criminals as informants, which was described in the *Tactics* section. Also note an article which appeared in *The Chicago Sun-Times* on May 31, 1998, entitled, *Family of Spies on Run*, which described an, "unknown number" of entire families who are citizen informants. "They are ordinary citizens who, for reasons ranging from petty revenge to pure patriotism, agree to work with the bureau," stated the *Times*. The article revealed that these families work with the FBI, DEA, CIA, RCMP, as well as local law enforcement. Most of these agencies are now part of DHS. "Because this is the lifestyle they have known almost since their birth, the ... children rarely complain," they wrote.

<u>Secrecy</u>

To my knowledge, at this time, no one has openly admitted to being part of this program. Targeted people have been stonewalled when confronting their tormentors in public. The citizen squads are obviously told that under no circumstances are they to admit they're part of an organized program to remove troublesome individuals from the community.

In Russia, the state emphasized the need for total secrecy when recruiting people into the Citizen's Brigade. According to Kourdakov, during their lectures they were told that, "Under no condition was the public to know what was going on." The reason was that, "some people could misunderstand what we're doing and why we have to do it," and "don't appreciate the danger that these people represent to our society." When citizen informants in East Germany were recruited, they agreed to a, "code of secrecy," and some had to sign a waver, which stated that they would not mention anything about their connection to the MfS.

By their behavior it is apparent that participants in the current program are told that targeted people can't be reasoned with, but must leave on their own accord, by way of a collaborative community effort. In all likelihood, stigma is probably used to justify the harassment. This includes labeling people threats to "national security," or criminals who have escaped justice.

These tactics have been used in other countries to justify persecution of political dissent. As part of their training, it is also possible that they engage in skits where they witness people breaking down or fainting in front of them, as a result of being systematically Mobbed out in public. Some of the training is apparently designed to put their conscience aside when harassing targets. As we'll learn in the next chapter, when you combine these desensitizing drills with lies, you can get most people to do just about anything.

"As long as people believe in absurdities they will continue to commit atrocities."
-Voltaire

Filtering

Another reason participants are told to deny their involvement in this program is obviously because their handlers know that because they operate on *huge lies*, they could not rationally justify the cause. It is critical that participants receive no outside information that can penetrate this barrier. Precautionary measures must have been taken to filter information that would undermine the group objectives.

It has been said that those who do not learn from history are condemned to repeat it. The people who created this program are obviously aware that it is *absolutely essential* that no parallel be made between similar historical events and what is taking place now. If this program originated from a Think Tank, you can bet that they have accounted for this, and it has been integrated into the training curriculum.

Therefore, certain phrases, such as "New World Order," "Dictatorship," "Cult," "Nazi Germany," "Hitler," "Cointelpro," MKULTRA," "Experimentation," "Gang Stalking," "Harassment," etc, are probably covered with mental *mindguards* to disengage critical thought.

Similarly, according to cult expert Steven Hassan's book, *Releasing the Bonds: Empowering People to Think for Themselves*, these mental blocking techniques are given to members of mind-control cults to reject information that undermines the cult. It basically means that when they're confronted with a particular phrase or question, they should think or say a certain phrase to reject it. This is used to stop critical thought and funnel external input into a mental recycle bin. It works remarkably well.

In his book, *Vital Lies, Simple Truths*, Dr. Daniel Goleman stated, "A mindguard is an attention bodyguard, standing vigilant to protect the group not from physical assault, but from an attack by information." "'A mindguard,' says Janis [author of Victims of Groupthink], protects the group 'from thoughts [information] that might damage their confidence in the soundness of the policies to which they are committed.'" More on the *Group Mind* will be covered in the next chapter.

The Nazis and their educated followers were essentially a *political cult*. The cult was fed massive lies which were reinforced by the careful filtering of information and propaganda conveyed by mainstream outlets. Literally, a whole country was under this spell. This is happening now. "Hitler and his troop of amateur magicians and hypnotists in the 1930's and 1940's were able to reconstruct a technologically advanced nation such as Germany into a vast theatre of illusion," stated Professor Marrs. "They were also able to induce educated and intellectual [people] into becoming ... oppressors of their fellow men..."

Some may disagree, but arguably, this international group is a political/corporate cult. Various cult experts themselves don't seem to agree on any set definition. But in my view this group is a cult. More evidence will be provided in the *Structure, Purpose*, and *Motivational Factors* chapters to illustrate that this group is an essential part of a massive political movement.

Communication

There is communication between the surveillance faction and the citizen informant squads and local businesses in the area. Obviously, the managers of local stores are contacted by a representative of this program, possibly by cell phone, pager, or walk-in, and told to prepare for the arrival of the targeted person. A picture or video of the targeted person may be shown to the participating staff.

The targeted individual may also be identified as he or she walks into the store. Consider that the IMs in East Germany communicated using non-verbal cues, so this is also a possibility. When the TI arrives, those who have been co-opted follow through with the tactics they've learned from their training. They also work in unison with plain-clothed citizens to harass the targeted customer. Advanced electronic methods of communication, which are not commercially available, may be used by the coordinating faction to direct specific citizen informants.

More than likely the current recruitment of informants is done using a beneficial appearing program, equipped with convincing propaganda, which is delivered by trusted authority figures. The recruitment programs offered to the public probably don't refer to this activity as "harassment," or "Gang Stalking."

In all likelihood it is cloaked in a beneficial sounding name. The behavior exhibited by these citizens indicates that they are absolutely convinced that this policy is legitimate and necessary. If it originates from a government Think Tank, you can expect no less than an intellectual and emotional masterpiece that is very cleverly delivered. This program was very well thought out.

Friends and Family

Friends and family of targeted people are sometimes used to harass them. As shocking as this sounds, identical tactics were used during Cointelpro. According to the book, *War at Home*, by attorney Brian Glick, FBI records reveal frequent maneuvers to generate tension (harassment) on targeted people by recruiting their "parents, children, spouses, landlords, employers, college administrators, church superiors," and others, into investigations. The word "frequent" leads me to conclude that, just as in East Germany, this was standard practice.

This helps to explain the current participation of friends, and family of targeted people. Moret added, "What we've studied and reported and identified is that neighbors are contacted and co-opted, [and] members of the target's family are co-opted..." Similarly, according to Wolf, the residences of targeted people were "put under siege," by the Stasi, and every family member, and every visitor, was kept under observation.

The recruitment of friends and family of targets may not just serve to harass, confuse, and isolate, but also as a discrediting strategy. After all, a person's friends and family would never systematically harass them, so someone complaining of such an ordeal must be mentally ill. This may be compounded if a handler forces a targeted person's family to have the target institutionalized. Remember, one in ten citizens will be recruited as informants anyway, so co-opting ten or fifty friends/relatives of a targeted person is not difficult.

How can a good friend or relative be recruited into a harassment campaign? Simple. They can be lied to, intimidated, or blackmailed into becoming informants during a bogus investigation. Then they can be given a gag order and threatened with jail time if they mention anything about it. Both the Stasi and FBI have done this. According to attorney Glick, blackmail and threats, using statements such as, "We know what you have been doing, but if you cooperate it will be all right," were often used as leverage for the recruitment of a relative or associate during Cointelpro.

Family members or associates who targets have a shaky relationship with can be bribed if they are greedy or in financial ruin. Naive ones or youngsters can be lied to by appealing to their sense of patriotism. This would be especially effective if delivered by a trusted community authority figure, such as a local official. When gaining cooperation in this way, these friends/family members may be fed some truth, and then some damaging lies can be thrown in.

After the friends and family of a targeted person have been recruited, they can be directed by their handlers to perform carefully scripted harassment skits, which include some mentioned in the Tactics chapter. This harassment by proxy is optimized by the use of sophisticated electronic surveillance equipment.

Summary

Multiple citizen informant programs are now operational, with the goal of recruiting a significant portion of the population. We've seen the political *Left* and *Right* justifiably compare these informant programs to ones that emerged in Germany and Russia. These programs are supervised by the Department of Homeland Security. They are similar in operation and purpose to those used in Russian and German dictatorships, which were created by the Committee for State Security and Ministry for State Security, respectively.

Those mentioned to be recruited, reflects (to some degree), those reportedly Gang Stalking targets. A massive network of citizens (probably millions worldwide), stalk and harass targeted people in public. They include a wide strata of individuals ranging from seniors to children, and often the friends and family of targeted people. Local governments are fully complicit in the harassment.

The federal government can also recruit groups or individuals as informants, which are not connected to any official organization, such as a Neighborhood Watch network. Without even being aware of the existence of the other network, these entities too can be used to harass people in public.

Endnotes

* Although the portion of the People's Brigade which Kourdakov wrote about didn't exhibit the exact traits as the current global stalking group (due to the frequent use of overt physical violence), his account does illustrate the emphasis on secrecy with state-run harassment programs. Furthermore, the Soviet Union did (and still does) possess a huge network of plain-clothed citizen informants, which is used to stalk (shadow) enemies of the state. See the report entitled, *Understanding the Solzhenitsyn Affair: Dissent and its Control in the USSR*, edited by the former Deputy Director of the CIA, Ray S. Cline. The report was based on a conference held by the Center for Strategic and International Studies (CSIS), at Georgetown University.

** On multiple occasions I have noticed children being used in skits which were obviously intended to convey sexual innuendos.

Chapter 19

Structure

Full Government Complicity

Reportedly, the ground crews that participate in patrols (Gang Stalking) are connected to academic institutions, the Department of Energy (DOE), the Department of Justice (DOJ), and other federal or state agencies. If this is the case, then this program is similar to the harassment campaigns in Germany which were run by the Stasi. They operated a vast network of citizen informants. Some of these informants were directed by the Stasi to conduct patrols, where they would systematically Gang Stalk enemies of the state.

Anna Funder described the Stasi as, "the internal army by which the government kept control." John O. Koehler, who served as a correspondent for the *Associated Press* for 28 years, says they were "a criminal organization founded and operated by a monopolistic party ... that trampled on basic human rights and on the dignity of the individual." The Stasi existed to protect the state from the people. More specifically, it existed to keep those in charge of the dictatorship in control.

Fulbrook wrote, "a key task of the Stasi was to ensure that the potential bases for collective action in opposition to ... the state ... did not develop," and that, "individual 'trouble-makers' or 'ringleaders' [were] rapidly isolated and disciplined." To accomplish this, they used their own version of Cointelpro (*Zersetzung*), which had the full support of all major components of the state, including a network of citizens which would stalk and harass people. The Stasi and the IMs were essentially the private armies of the individuals who controlled the country.

Apparently these harassment campaigns can legally be done. Attorney Brian Glick wrote, "Much of what was done outside the law under Cointelpro has since been legalized by Executive Order NO. 12333," which allows "counterintelligence activities" by the FBI, CIA, the military, or anyone acting on their behalf. "Specialized equipment," and expert technical assistance may be used "to support local law enforcement." All are free to mount electronic "surveillance" and conduct "break-ins" without a warrant. This means that any federal or local law enforcement agency or government contractor can conduct these "investigations." These legal "break-ins" account for the Gaslighting tactics that targets experience.

It would be necessary for "the FCC as a minimum and the Department of Energy as a minimum to have some oversight and control over what is going on," stated McKinney. "[Each] government obviously is complicit," she declared, "because otherwise these ... operations would not be allowed to exist." McKinney's findings suggest that the FBI, CIA, and DOE are probably involved in these harassment campaigns.

Moret added, "the military is ... very directly and heavily involved in it. But it's not just the military," she says, "it's the CIA, FBI" and other "law enforcement" agencies. "This globally infiltrated organization," declared Dr. Kilde, "has "octopus type" activities in all major intelligence services in the world," and has "recruited people from all important government institutions, state and local administrations."

The April/May 1996 issue of *Nexus Magazine* provided us with another clue with the publication of an article entitled, *How the NSA Harasses Thousands of Law Abiding Americans Daily by the Usage Of Remote Neural Monitoring (RNM).* The article stated, "The NSA gathers information on U.S. citizens who might be of interest to any of the over 50,000 NSA agents (HUMINT). These agents are authorized by executive order to spy on anyone. The NSA has a permanent National Security Anti-Terrorist surveillance network in place."

The Big Lie

The elite have a tendency to fragment organizations within their hierarchy in order to keep people ignorant as to how their organization is a small component within a larger operation. This is how an organization with seemingly positive objectives can be used for destructive means. Therefore, the Hidden Evil is probably highly compartmentalized and the various factions within it are almost certainly unaware of the existence of one and other. For instance, the citizens who operate within the public front are probably not aware of the use of Directed Energy Weapons, murders, or framings. Neither are they likely to be aware of the real agenda.

An analysis of the tactics and strategies used indicates that there must be at least several factions to this highly organized program. There is a recruitment and training unit, which provides training to individuals, businesses, and groups within communities. This faction more than likely appears professional and is equipped with supporting propaganda delivered by trusted authority figures, who claim that the program is a matter of "national security." The individuals in this unit are probably operating on lies themselves, because in order to *sell* such a program, you first need to convince the sellers.

There is a logistics faction which coordinates public harassment such as the arrangement of people, and vehicles, as well as informing federal, state and local entities of a targeted person's pending arrival at their enterprise. There is an intelligence faction, which puts people under surveillance. The surveillance is apparently justified by a bogus investigation. Some of these factions may be merged into a single team. For instance, the logistics faction is obviously in direct contact with the surveillance faction, and they may even be the same team. I refer to the combination of these factions as, the *public front*, because in all likelihood, it appears beneficial and innocent. The Citizen Corps National Neighborhood Watch program, figures prominently.

According to the Citizen Corps' *Guide for Local Officials*, each community has a Citizen Corps Council composed of representatives from law enforcement, fire and EMT services, faith-based groups, community-based groups, elderly and minority populations, transportation systems, utility companies, businesses, educational institutions, environmental groups, the media, and "[any] groups that represent a large cross-section of the community." Basically they recruit people from all of these major groups within the community.

The guide explains, "At the state level, all Governors have appointed a state coordinator for Citizen Corps to facilitate this locally driven initiative." Notice the operative word, "all." Elected officials such as mayors and council members are also cooperating. "The state coordinator," it continues, "will work closely with the local governments, other state organizations, the Department of Homeland Security, (DHS), the Department of Justice (DOJ), the Department of Health and Human Services (HHS), and other federal agencies to implement a successful Citizen Corps program for the state."

This means that your high ranking elected officials, as well as federal and local law enforcement are *fully* aware of this program, and are donating significant resources to it. The Citizen Corps' Neighborhood Watch program is essentially, according to the NSA, the front line for homeland security.

So, if this program is in fact the public front that is used for mass recruitment, then it would explain many things, such as:

- The colossal network of volunteers (informants) conducing citizen patrols (Gang Stalking), which is directed by FEMA and the DHS, working with local police.
- The near-perfect coordination of non-stop harassment, which unfolds as targets travel around town, including any public place, store, or restaurant.
- The tremendous state resources that are used which obviously require much time, effort, and financial backing.
- The stalking of people by vehicles representing entities such as, the U.S. postal service, UPS and FedEx, police cars, fire trucks, ambulances, local business, utility companies, city vehicles, taxis, and public transportation vehicles. And in some cases, aircraft such as small propeller planes and helicopters.
- The persecution of targeted people inside essentially any local, state, or federal establishment, by employees who work with citizens on patrol to harass them.
- The "all hands on deck" philosophy of the entire state engaging in well-orchestrated harassment.
- The participation of a wide-range of citizens, such as seniors groups, school children, religious and ethnic organizations, etc., which are presumably drawn from various groups within the community.
- Because training originates from the same source, it would explain the identical tactics used nationally—even globally.
- The cover-up by local and federal law enforcement, as well as agencies which exist to help people being tortured or persecuted.
- Because it is a DHS initiative, it would explain why no organization has stopped it.

What the Public Doesn't See

Citizen Corps seems to fit the observable operational blueprint previously noted. It appears to be a likely candidate for the public front. But there are definitely other factions, which are probably unknown to the majority of the participating public. Similar to the public front, some of these factions may be combined into a single unit.

These factions are also obviously directed by federal and local law enforcement.

They include, a faction for co-opting specific individuals in a targeted person's life (friends and family), which appears to involve lies and intimidation. There is a faction which attacks people with Directed Energy Weapons. There is a faction that conducts break-ins, vicious rumor campaigns, staged accidents, framings, and blacklisting.

The program is "very hierarchical" Moret describes. She says there is a lower faction consisting of criminals who are "already compromised" due to their participation in illegal activities. These criminals participate in order to gain recognition with local officials and other figureheads. According to Dr. Kilde, this recruitment also includes the use of the mafia and terrorists when necessary. Interestingly enough, it's a matter of public record that the FBI works with known criminals, including those in the mafia who are confirmed murderers, to assist with framing innocent people for crimes. Apparently this is done for reasons of "national security." It's also a matter of public record that the U.S. Government has funded, trained, and even protected terrorist networks such as Al-Qaeda.

Secret Police

"Secret police (sometimes political police) are a police organization which operates in secrecy for the purpose of maintaining national security against internal threats to the state. Secret police forces are typically associated with totalitarian regimes. As their activities are not transparent to the public, their primary purpose is to maintain the political power of the state... A state with a significant level of secret police activity is sometimes known as a police state."
-Wikipedia Encyclopedia

To recruit regular citizens to perform nefarious acts such staged accidents and framings, these officers can scan for certain qualities in an informant database consisting of *literally millions* of people, in order to select the mentally ill, criminals, or those who can be blackmailed. Then, these informants, who have no official ties to the government, can be co-opted into a staged event for reasons of national security.

In this manner, these factions may feed a steady supply of innocent people to the public front, in order to promote group cohesiveness—thus ensuring the survival of the program. These factions represent the true underlying psychopathic nature of the program. Based on my research, the public front is simply a *mask* used to recruit the public into a very cleverly disguised program of torture and murder—The Hidden Evil.

Ted L. Gunderson, a retired FBI Senior Special Agent insists there are two types of investigative operations in the U.S. intelligence agencies, overt and covert. He says, "The overt operations involve investigators who respond to various federal violations of the law such as bank robbery, kidnapping, extortion, etc..." But the "covert operation" says Gunderson "involves those who never identify themselves and are involved in harassing and targeting citizens..."

So, if Citizen Corps is the current major operational vehicle for this program, then FEMA, federal and local law enforcement (with full support of the entire community), are directing these persecution campaigns on the local level. And this unseen unit, which operates beyond the public front, is responsible for conducting, framings, murders, staged accidents, torturing people with Directed Energy Weapons, blacklisting, and mocking people with V2K—all done under the banner of national security.

Who created the Hidden Evil? The creators of this program have enough global influence to initiate policy in all NATO nations. They also have the authority to issue worldwide stand-down orders on organizations like the Red Cross, ACLU, Amnesty International, etc. And they have the money to finance it.

Who benefits from this program? When investigating a matter, if the evidence you've been able to gather indicates a specific pattern which is not easily produced, then it is reasonable to conclude that any entity which can produce the pattern, and has done so in the past, is suspect. Furthermore, it is also reasonable to conclude that if the same entity has the motive, the means, and the opportunity, then it is suspect.

Hypothesis A

In his book, *America's Secret Establishment*, Professor Antony C. Sutton describes a hypothesis as "a theory ... which has to be supported by evidence." He continues, "Now in scientific methodology a hypothesis can be proven. It cannot be disproven (sic). It is up to the reader to decide whether the evidence presented later supports ... the hypothesis." Sutton states that no reader can decide until they've absorbed all of the evidence presented.

Think Tanks created the Hidden Evil. This includes, but is not limited to, The Bilderbergers, The Trilateral Commission, the Council on Foreign Relations, and their interlocks, funded by the Tax-exempt

Complex.(*) These organizations are heavily interlocked with Wall Street and the Federal Reserve. Policy set by these unelected rulers becomes law with regularity. The Hidden Evil was not a willful creation of the public that voted it into existence, it was done by secrecy. Essentially, it is the creation of what congress, federal law enforcement, and researchers have called, the *Invisible Government*, or the *Shadow Government*.

The following three individuals have worked with government Think Tanks, and have evidently contributed to the fine-tuning of this program. They include Dr. John Alexander, Janet Morris and Chris Morris.(**) The Morris' and Dr. Alexander have worked with Think Tanks such as the Council on Foreign Relations, the Center for Strategic and International Studies (CSIS), and the U.S. Global Strategy Council (USGSC) researching Non-lethal Weapons technologies. They have also instructed government officials on the use of Neuro-Linguistic Programming (NLP). Here is a brief biography of each, which was taken from their websites.

Mr. Chris Morris

Mr. Morris has worked with organizations such as Raytheon, Alliant Techsystems, Olin Ordnance, Primex, ARDEC, Delfin Systems, the Millburn Corporation, the Human Potential Foundation, Lawrence Livermore National Lab, Argonne National Lab, Penn State Advanced Research Lab, Westinghouse Electronic Systems, BDM International, the United Kingdom Ministry of Defense, USGSC, CSIS and others.

Prior to forming M2 Technologies with his wife Janet, Mr. Morris was "Research Director at the U.S. Global Strategy Council [USGSC]," which he describes as a "Washington-based think tank" founded by Ray S. Cline, the former Deputy Director of the CIA. According to Mr. Morris, this Think Tank was composed of "a select panel of senior government and industry experts who met regularly to formulate and guide the Council's non-lethality agenda." He briefed "executive branch, DoD, and Congressional officials on non-lethal concepts, technologies and strategy."

Mr. Morris states that he has received "multiple grants to study non-lethal technology issues for the MHW Foundation and the Winston Foundation." He adds that he has "participated in defense technology roundtables of the Heritage Foundation."

Ms. Janet Morris

Ms. Morris is President and CEO of M2 Technologies, Inc. (M2), which specializes in "non-lethal weapons (NLW), novel technology applications, tactics and technology." She has taught or provided course material to the U.S. Air Force's Air Command and Staff College, National Defense University's (IRMC) School of Information Warfare and Strategy, and Penn State University Applied Research Laboratory's Non-Lethal Institute.

She has provided and presented seminars and briefings to the Defense Science Board, the Congressional Research Service, Senate Armed Services Committee Staff, and the Center for Naval Analysis. Like her husband, Ms. Morris worked at the USGSC as Research Director of Non-lethals, where she provided "strategic planning support to U.S. Government agencies, departments, and Congressional offices."

According to McKinney, in their publication entitled, *Nonlethality: Development of a National Policy and Employing Nonlethal Means in a New Strategic Era*, the USGSC outlined the foreign and domestic use of Non-lethals, and made numerous references to the domestic use of this technology against elusive "enemies" of the U.S. Government.

"The term, 'non-lethal,' used to describe this technology is misleading," says McKinney. "The energy emitted from all of these [Microwave] weapons can kill when appropriately amplified. At lower levels of amplification, they can cause extreme forms of physical discomfort and debilitation."

The Morris' have provided interviews and opinion pieces for *The Smithsonian*, *Newsweek*, *The Wall Street Journal*, *BBC*, *ABC*, *Discovery Channel*, *The Learning Channel*, and other media. According to the *New York Times Magazine*, both have worked on a task force for the Council on Foreign Relations.

Col. John B. Alexander, PhD

His biography reads: "Dr. John Alexander has been a leading advocate for the development of non-lethal weapons..." He "organized and chaired the first five major conferences on non-lethal warfare, and served as a US delegate to four NATO studies on the topic." Dr. Alexander has worked with the "highest levels of government," including the White House, NSC, CIA, and other senior defense officials.

He has written articles for *Harvard International Review*, *The Boston Globe*, *The Futurist*, *The Washington Post*, and other media.

He has appeared on *Fox News*, *Larry King*, *CNN*, *Dateline* and international television. Currently he works as president of LEADS, Inc., and serves as an adviser to the Commander of U.S. Special Operations Command.

He was "instrumental in influencing the report that is credited with causing the Department of Defense to create a formal Non-Lethal Weapons Policy." According to Dr. Alexander, the creation of this policy resulted from his non-lethals study for the Council on Foreign Relations.

An article entitled, *The Quest for the Nonkiller App*, which appeared in the July 25 issue of *The New York Times Magazine*, stated that Raytheon had developed the Active Denial System (ADS) which causes "palpable pain" and "can operate beyond small-arms range, enabling an operator to deter a foe long before a potentially fatal clash occurs." According to the article, Stephen Goose, Director of the Arms Division of Human Rights Watch, "paints an Orwellian picture in which repressive regimes obtain nonlethal weapons to keep restive populations in check without resorting to the sort of bloodshed that can earn a country unwanted attention." That is exactly what is happening now.

The article continued, "Janet Morris of M2 Technologies would like to see, 'calmative agents'—weaponized versions of Valium and other drugs—deployed in battle." The Council on Foreign Relations has issued reports and recommended an increase in research finances, it said, and added, both "[Janet and Chris Morris] have worked as members of the Council on Foreign Relations' task force."

Summary

Experts in Neuro-linguistic Programming (NLP), and Non-lethals have worked with elite Think Tanks that have been known to control most governments. These Think Tanks are advocating the creation of a single world government with them in complete control. The Non-lethals which they have sponsored are reportedly used on people worldwide. NLP is allegedly used as a weapon during citizens' watch patrols (Gang Stalking).

Endnotes

* Although there are probably entities coordinating these Think Tanks, they exist at the upper echelon of this control structure and are a recognizable element of it.

** These are not the only individuals who helped create this. But they are an identifiable link between the government Think Tanks, the non-lethal technology, and NLP, which are used in this program.

Chapter 20

Purpose

New Weapons for a New World Order

Dr. Alexander and the Morris' have made multiple references to the term, "New World Order" in their publications. It is unclear whether they are aware of what this New World Order will be: the history of some of the organizations which they have served, or their true objectives.

USA Today ran an article entitled, *Top Military Advisor Signs on for Video Games*, on July 3, 2003, which stated that Dr. Alexander has "inked a contract ... with Platinum Studios." "We're talking about everything from electromagnetic pulse weapons that could cripple a country by wiping out all its electronics, to controlling insects with pheromones to make whole cities uninhabitable" says Platinum Studios Chairman Scott Mitchell Rosenberg." It continued, quoting Dr. Alexander who stated, "This stuff is real ... the genetic and biological capabilities that exist right now, the cutting-edge weapons, technologies and threats—these are real scenarios that we could literally face today, and it's mind-blowing."

Market Wire ran an article in March, 2004, entitled, *Colonel John B. Alexander, Ph.D. Secured as ShockRounds (TM) Advisor.* "Colonel Alexander" it explained, "is recognized and acknowledged as the world's leading authority on non-lethal weapons and is credited with developing the modern concept of non-lethal defense," and "continues to consult to the US government and provide guidance in military situations globally ... on non-lethal weapons."

In an article entitled, *Mind Games*, by *The Washington Post* on January 14, 2007, Dr. Alexander stated, "Maybe I can fix you, or electronically neuter you, so it's safe to release you into society ... It's only a matter of time before technology allows that scenario to come true..." The program is already here and it is nothing less than an electronic concentration camp.

Dr. Alexander authored an article called, *New Weapons for a New World Order*, which appeared in the March 7, 1993 issue of *The Boston Globe*. He wrote "[The] United States must be able to protect national interests and values, even in ambiguous circumstances." And that more options will be required in the future to deal with "threats to national security." Remember, the Morris' and Dr. Alexander have suggested the *domestic use* for these weapons. What this means in plain English is that opponents of big corporations, or anyone against any element of the New World Order are the enemies.

Former FBI Agent Dan Smoot wrote, "Through many interlocking organizations, the Council on Foreign Relations 'educates' the public—and brings pressures upon congress—to support CFR policies. All organizations, in this incredible propaganda web, work in their own way toward the objective," which Smoot contends is a one-world socialist dictatorship. Allen warned that under the control of the Tax-exempt Foundations which are "all extensively interlocked with the C.F.R." an array of scientists such as, "geographers, psychologists and behavioral scientists to natural scientists, biologists, biochemists and agronomists are making plans to control people."

In February 1995, *Wired Magazine* explained, "Janet Reno and Deputy Secretary of Defense John Deutch last April signed a 'memo of understanding' under which the Pentagon will share 'dual-use' nonlethal technology with domestic law enforcement agencies. ... Alexander and members of the Council of Foreign Relations, an independent advisory group, met with then-Defense Undersecretary John Deutch on the pacifying potential of non-lethals." "The entire concept of national security has changed," stated Dr. Alexander. The article described Dr. Alexander as "'the intellectual leader in the field' according to Robert Kupperman, a senior advisor at the Center for Strategic and International Studies [CSIS], a Washington, DC, think tank."

It continued, "Traditional US military doctrine—using overwhelming force to 'break things and kill people'—has its limits in the New World Order..." "[Janet] Morris, a research director of a DC think tank allied the US Global Strategy Council, and her husband, Chris, a fellow writer and national security expert, were also intrigued by the nonlethal possibilities of some of the futuristic technologies being researched at the government labs." Finally, it stated, "In *Nonlethality: A Global Strategy*, she [Janet Morris] trumpeted the concept as a 'revolutionary strategic doctrine' that would allow the US to seize the moral high ground in coping with the demands of a New World Order."

In an article entitled, *Optional Lethality*, from *Harvard International Review*, Dr. Alexander stated, "Finally, there is a small but vocal group of conspiracy theorists that view non-lethal weapons as tools for illegally controlling the civilian populace. This argument fails the test of logic." He sites one example in which a riot control agent is used to disperse a crowd, and the how the media "equates that application to the use of lethal gas in the Nazi death camps." He explained, "It is the people who use technologies for evil purposes..."

I completely agree with his last sentence. But I wonder if Dr. Alexander realizes that the interests he serves, such as the CFR, have been linked to at least one attempt to overthrow the U.S. Government by force, as well as the funding of Nazis and Communists. Additionally, the torture and persecution of innocent people was made *legal* by legislation during German and Russian dictatorships. This is standard procedure.

On February 23, 1954, Senator William Jenner warned during a speech, "Today the path to total dictatorship in the United States can be laid by strictly legal means, unseen and unheard by the Congress, the President, or the people. ... Outwardly we have a Constitutional government. We have operating within our government and political system, another body representing another form of government..."

Hypothesis B

The Hidden Evil is being used by the financial elite to covertly destroy anyone they choose as part of a political movement toward a worldwide tyrannical dictatorship.

The evidence suggests that the primary purpose for the existence of this program is the same as those which existed in Russia and Germany. The evidence also indicates that people are targeted in order to be covertly murdered, so they may be recruited, influenced to commit acts of violence such as "going postal," or driven to suicide. This is accomplished using not only the informant squads, but also an assortment of tactics outlined in the *Tactics* section. Due to heavy compartmentalization, these informant squads and others who participate in this program are, in all likelihood, not aware of its true purpose or of its other factions.

"[A] number of virtually identical complaints have also been received from England, Canada and Australia," wrote McKinney. "It would appear that these activities are also ongoing on the European Continent, and the former Soviet Union." She explained, "Connections between this global pattern of activity and the objectives of the so-called New World Order remain to be determined." From the information I've come across, there is no doubt that this program is part of the New World Order political movement.

This program is being used to silently neutralize (neuter), all people that the Establishment believes will be troublesome to their rule. This usually includes freethinking, incorruptible, nonconformists. This movement toward a one-world government (dictatorship) has been called the *New World Order, Global Union*, and by those who know what it will be like, *The 4th Reich*. Professor Marrs advised, "The awful truth is that the entire democratic system—euphemistically called The New World Order—now being set up and organized ... is to be a dictatorship of the worst kind." By now you already know why the mainstream media has not informed you of this impending tyranny.

As revealed by congressional investigations, the Tax-exempt Foundations have been known to promote socialism and one-world government. "The major foundation complex," says Wormser, has operated as an "integral arm of the government." He concludes that "If a revolution has indeed been accomplished in the United States, we can look here for its ... impetus." They have been known to fund the social sciences (behavior modification and manipulation). At least one has funded mind-control experiments, which appear to have contributed to the creation of the Hidden Evil.

There is nothing new here. This is the same pattern repeating itself. In order to establish control of a country (or planet) you must create a threat. When these harassment programs started out in other countries such as Russia or Germany, they began with a positive intent. They always mask these state-run programs by targeting undesirables and terrorists—only! But because their true purpose is to destroy opposition, they *always* expand.

Then they move on to groups of people that do not approve of their corruption. They demonize these non-criminal groups and individuals by linking them with a previously accepted undesirable group. Those in control of state encourage the harassment of people. They get the children involved, as an "all hands" approach to "help out the community." Remember that in the past, on the local dispensing end, these programs had the complete approval of leaders of the communities. In other words, these programs were approved and carried out by *the good guys*.

The evidence so far suggests that the United States and other NATO countries are, functionally, covert socialist states with a democratic/republic frontage. The evidence also suggests that this "front" will be replaced by an overt tyrannical socialist dictatorship. Although we've been led to believe that Communism and Fascism are opposites, both are on the same end of the spectrum, which is a socialist totalitarian rule. Socialism may sound beneficial on the surface to some, as it promotes an even distribution of wealth, so that everybody gets a fare share.

However, in practice it is much different. It consists of a military/police state, heavy surveillance, and concentration camps. Germany and Russia are prime historical examples. Allen explained, "Here in the reality of socialism you have a tiny oligarchial clique at the top, usually numbering no more than three percent of the total population, controlling the total wealth, total production and the very lives of the other ninety-seven percent."

"The dignity, liberty, and rights of the human being are protected by the criminal laws of the socialist state. The socialist society is guided by the respect for human dignity, even ... the violator of the law, that is the steadfast mandate for the activity of the state and of justice."
-Article 4 of the Constitution of former East Germany

"Totalitarianism under any label spells loss of freedom," wrote Sutton and Wood. Senator Barry Goldwater charged, "In their pursuit of a new world order they are prepared to deal without prejudice with a communist state, a socialist state, a democratic state, monarchy, oligarchy—it's all the same to them." "If you have total government," wrote Allen "it makes little difference whether you call it Communism, Fascism, Socialism, Caesarism, or Pharoahism ... [because it's] all pretty much the same from the standpoint of the people who must live and suffer under it."

Now it becomes clear why the elite monopolists fund both Communism and Fascism. They are a consolidation tool used by the elite to install dictatorships. "The key to modern history," wrote Professor Sutton, "is in these facts: that elitists have close working relations with both Marxists and Nazis."

Allen agreed, writing, "If one understands that socialism is not a share-the-wealth program, but is in reality a method to consolidate and control the wealth, then the seeming paradox of super-rich men promoting socialism becomes no paradox at all. Instead it becomes the logical, even the perfect tool of the power-seeking megalomaniacs." He reveals, "Socialism, is not a movement of the downtrodden masses, but of the economic elite."

So according to these researchers the elite use socialism as a consolidation and control mechanism. Allen describes Socialism as "a movement created, manipulated and used by power-seeking billionaires in order to gain control over the world." He points out that this is accomplished, "first by establishing socialist governments in the various nations" and then combining them all under the "United Nations." Socialism, in practice, is *tyranny*. It is *control* by the wealthy individuals who control the state. It is always oppressive, because they must use terror of some kind to create restrictive laws which ensure their control.

Senator Goldwater sited an example of how the Communist Revolution "was not an attempt to destroy the state but rather to seize control of the political state," and to "replace the individuals exercising ... power." What these researchers are saying is that there has been a concerted covert effort by the financial elite to overthrow the U.S. and other NATO governments and replace them with puppet governments. The evidence suggests that for the most part, it has already been done.

The Revolution is Over

This was not a takeover where the flag was captured and destroyed, but a covert takeover from within. The evidence for this was presented in Volume I. As long as they can vote every four years to decide which elite-controlled puppet will be installed in the White House, most people seem to be content. This was a covert revolution, done by controlling the money supply, legislation, and infiltration, rather than overt bloodshed.

"The most important thing of all is to remember that the political coup de grace is over—the virtual domination of the White House," stated Sutton and Wood. Evidently they found it necessary to leave the flag, the stature of liberty and the White House intact to keep people convinced they're still in America.

Sutton and Wood describe it as a, "socialist revolution by stealth, rather than by blood in the streets, but revolution just the same." A "Totalitarian Takeover of the U.S. Executive Branch" has already happened they explain. "A straightforward and reasonable conclusion is that there has apparently been a covert fascist (national socialist) takeover of the United States government. By fascist we mean a corporate socialist state..." Wormser agreed, writing, "All that can be hoped for is a counterrevolution."

"Revolution is always recorded as a spontaneous event by the politically or economically deprived against an autocratic state," wrote Professor Sutton. He adds, "Never in Western textbooks will you find the evidence that revolutions need finance and the source of the finance in many cases traces back to Wall Street." If people knew that their country had been stolen they would be outraged. It is essential that they believe they're free and informed, as the elite cannot control the whole population physically.

One critical element for the creation and maintenance of a dictatorship is the implementation of a massive network of citizen informants. Both Germany and Russia had them. The "IMs were to play an important role in the Stasi's policy of 'subversion,'" wrote Catherine Epstein in her publication, *The Stasi: New Research on the East German Ministry of State Security*. "They had made possible the regime's assault on privacy, honesty, and truth." "In its use of such a large network of secret informants," she commented, "the MfS was surely imitating Soviet practice."

The CSIS report explained that the control mechanisms in Russia, which included the targeting of the civilian population, was only made possible by the "informant network which ... [was] the basis of KGB power." Referring to this report, McKinney added, "The KGB's success depended on the extensive use of informant networks and agents provocateurs..." She continued, "Participants in the conference agreed that the KGB's obvious intent was to divide and isolate the populace, to spread fear, and to silence dissenters."

In her book, *The People's State*, Mary Fulbrook described the former East German regime as a "participatory dictatorship" in the sense that a "very small ruling elite, with a linked apparatus of repression and injustice was supported and sustained by a very much larger number of people." She explained, "All key decision-making was concentrated at the top of the political pyramid." "The undoubtedly dictatorial political system was 'carried' by the active participation of many of its subjects," she added.

She continued, "The sinister Stasi, with its long shadow of unofficial informers (IMs), was everywhere, and observed and reported on everything. ... Anyone found to have transgressed permissible limits, in whatever way, could be subjected to brutal measures of repression ... [and were] subjected to sometimes unintelligible distortions and miseries in both professional and private lives." On June 16, 2003, in an article entitled, *Informant Nation*, *The New American* stated, "Every unfortunate society saddled with a police state has ... [used] citizen informants who play the most important role in enforcing conformity to the ruling elite's will."

Semantic deception is often used to disguise the true subversive intentions of those on the higher levels of the pyramid who operate the groups. Therefore, names such as "Peace Keeping Squads," "Citizens Brigades," "People's Police," "Crime Prevention Networks," and "Neighborhood Watch" may be used. The primary purpose for these groups and the accompanying DEW harassment is to *terrorize* any opposition to the financial elite's rule. The Hidden Evil is one foundational element for what will eventually become an overt dictatorship.

"Whoever can conquer the street will one day conquer the state, for every form of power politics and any dictatorship-run state has its roots in the street."
-Dr. Joseph Goebbels, Nazi Propaganda Minister

McKinney explained the reason for the existence of the current Gang Stalking program as "capturing a percentage of the population in order to install a dictatorship." Once more, this is essentially what happened in Russia and Germany by the same financial interests that now control North American and other NATO nations. She explained, "There is always a percentage of the population, roughly 20% or so, who will buckle ... and eagerly join the effort at destroying the remainder of the population."

"I'd like to go back to the underlying purpose for all of this," commented Moret, referring to this worldwide secret policy. "I'd like to read a couple of quotes. One of them is, 'We should crush in its birth the aristocracy of the moneyed corporations who already bid defiance to our government.' Those are the words of Thomas Jefferson."

"And another quote," she continued, "'I see in the near future a crisis approaching that unnerves me and causes me to tremble for the safety of my country. Corporations have been enthroned, an era of corruption will follow, and the moneyed power of the country will endeavor to prolong its reign, by working upon the prejudices of the people, until the wealth is aggregated in a few hands and the republic is destroyed.' Abraham Lincoln said that in 1862, and that is exactly what's happening."

McKinney warned, "I think that once full control is established over a major percentage of the population, and enough of the population is silent and unwilling to stick their necks out, that we inevitably would be heading toward a holocaust." In a chapter entitled "Revolution is nearly accomplished," Wormser stated, "This new revolution ... has become an attempt to institute the paternal state in which individual liberty is to be subordinated and forgotten..." He warned, "The system will inevitably call [for] .... the suppression of liberty and freedom," and that it "cannot exist without force and oppression."

The financial elite are fully aware that this secret policy is being used to hound and torture innocent people. This terror is critical to keep people under control who might be a threat to their rule, or people they just don't like. Allen warned, "In order to solidify their power in the United States they will need to do here the same thing they have done in other countries. They will establish and maintain their dictatorship through stark terror." He adds, "The terror does not end with the complete take-over of the Republic. Rather, then terror just begins."

Dan Smoot advised that the interlocking Think Tanks and Tax-Exempt Complex want "America to become part of a worldwide socialist dictatorship." As demonstrated, congressmen, presidents, federal law enforcement, whistleblowers and researchers, have all warned us of this. "If you want a national monopoly, you must control a national socialist government," wrote Allen. "If you want a worldwide monopoly, you must control a word socialist government." He concludes, "That is what the game is all about."

According to McKinney, there has been a progressive worldwide spread of this program over the years. Referring to targeted people, she stated, "I would say that the persons who have *realized* what's going on are just a drop in the bucket." Dr. Munzert recognized the consistency of reports in North America and Germany. He attested, "In many other countries," such as "England, France, Italy, Spain, The Netherlands, Switzerland, [and] Austria, there are also victims. It looks like they are all over Europe, and maybe all over the world." "I think the number of unknown cases could be much larger," he adds, because most people "cannot figure out what [is] ... really happening."

"When I was running the Electronic Surveillance Project I was in extensive correspondence with people overseas," declared McKinney. "The patterns were the same, the nature of the Gang Stalking and the harassment were the same." She says, "These are global operations," and described that the "patterns, the protocols, are virtually identical on a global scale." "It's very hierarchical," agreed Moret, "it's international; it's carried out in all of the NATO countries." She explained that this program is being, "widely used in the civilian sector in all of the NATO countries on targeted individuals who are thought to stand in the way of corporatism, capitalism, and fascism."

Earl's non-profit organization, An Ethical Light, has seen an increase in these activities as well. She states, "There is evidence that this is escalating, that there are more and more victims all the time." She says that her organization is hearing an increase in reports of "Gang Stalking," and "Street Theatre." "I've seen a tremendous expansion of these activities since the early 1990s," said McKinney, "and it has moved forward in [a] consistent fashion and become ever-more sophisticated and ever-more wide-spread."

Dr. Kilde warned, "They are ALREADY in every apartment block!" [Emphasis in original] This type of expansion is almost exactly like what occurred in Germany, where IMs were recruited from businesses, restaurants, prisons, factories, schools, sporting centers, hospitals, and workplaces. This was not accomplished by placing a couple of IMs inside these organizations. It was done by recruiting top management into a state-run program.

The squad leaders of these Neighborhood Watch groups, are referred to as, *Block Captains*. The official website for the city of Portland, Oregon, states that these individuals report directly to the police, fire, and other departments of the local government. It describes Block Captains as, "neighborhood leaders working to make their communities safe... By becoming a Block Captain, you can build and strengthen partnerships with neighbors, crime prevention, neighborhood associations, and the police bureau." "Most Block Captains," it adds, "work with their neighbors to organize a Neighborhood Watch, Business Watch, Apartment Watch, and a community foot patrol."

Funder emphasized, "There was someone reporting to the Stasi on their fellows and friends in every school, every factory, every apartment block..." Koehler agreed, writing, "Without exception, one tenant in every apartment building was designated as a watchdog [IM] reporting to an area representative of the Volkspolizei [people's police]."

These DHS-sponsored Neighborhood Watch groups are expanding rapidly. The Department of Justice proclaimed, "Law enforcement, in partnership with neighbors across the nation, have come together to meet the challenge and double the number of Neighborhood Watch groups... If you are not involved in Neighborhood Watch, [we] ... urge you to join." "As groups continue to grow," described the Citizen Corps program initiative, "the roles of citizens have become more multifaceted and tailored to local needs."

"The Neighborhood Watch Program," explained the Citizen Corps Guide for Local Officials "has been reinvigorated to increase the number of groups involved in crime prevention and homeland security." USAonWatch, the Citizen Corps focal point for all Neighborhood Watch groups, has on the top of its homepage, under the title, Register Your Watch Group, the words, "Recruit and Organize as many neighbors as possible."

Regarding the same pattern unfolding in Europe, Dr. Munzert attests, "It's strange, before we were attacked most of the victims were quite ordinary people who thought their human rights were certain." But after they discovered they were targeted and contacted law enforcement and government agencies, they were denied assistance. They assumed it was because these agencies were not aware of this program, but later found out they were. Dr. Munzert says they found this "absolutely strange" and couldn't understand why this was the case. He describes, "The same stories are heard about victims in other parts of Europe and in America."

What are these citizen informant squads? They're a portion of the population that it was necessary for the financial elite to recruit in order to generate a silent socialist revolution. These people are literally the instruments of their own oppression. Speaking of the elite's repeated use of citizens as leverage for a revolution, Allen describes them as "pawns, shills, puppets, and dupes for an oligarchy of elitist conspirators working about to turn America's limited government into an unlimited government with total control over our lives and property."

The program itself is part of a movement toward a socialist, one-world government. Resisting this "New World Order" will require the same "dedication and effort to win that it took to destroy Hitler," declared Allen. The elite have declared war on anyone who gets in their way during this movement. This includes people that their servants on the local level have identified as troublesome or resisters.

"The lie can be maintained only for such time as the State can shield the people from the political, economic and/or military consequences of the lie. It thus becomes vitally important for the State to use all of its powers to repress dissent, for the truth is the mortal enemy of the lie, and thus by extension, the truth becomes the greatest enemy of the state."
-Dr. Joseph Goebbels (Nazi Propaganda Minister)

Summary

The evidence presented so far (contained in Volume I and this chapter), suggests that the Invisible Government, consisting of an interlock of Think Tanks, foundations, Wall Street, and the Federal Reserve have taken over the true government that we learned about in our history books. As a matter public record, the promoters of Non-lethal Weapons, which have worked with these Think Tanks, have peddled the domestic use of these weapons against threats to national security, as part of the New World Order.

The information further implies that they are implementing a dictatorial model which was used in Germany and Russia, complete with a vast network of citizen informants. These groups participate in patrols (Gang Stalking), and other types of harassment on behalf of the state, which is ultimately controlled by people of wealth.

Chapter 21

Motivational Factors

This chapter will focus on the motivational factors for citizens on the street level only. The driving force for the upper level controlling faction of this movement will be explored in the chapter on psychopathy.

Intimidation and Blackmail

From what I've noticed, some of those who participate in this program do not do so of their own free will. They have obviously been forced. Sometimes, with their body language, they'll tell you, "I know, I'm sorry, it's not me, I have to do it," etc. Some of these people were previously targeted and know what will happen if they don't participate. Or, they willfully served and later found out what they were serving.

So, some do it out of fear. As demonstrated, both U.S. intelligence and the Stasi have recruited informants by intimidation. According to the Stasi's own figures, over 7% of the IMs were recruited by intimidation. Fulbrook states that this doesn't' take into account those who cooperated due to the implied retaliation for refusing to do so.

Only in the last ten years or so has information become widely available on this topic. Imagine that you're face with this horrible situation. Your life is falling apart. You know that something is happening, but don't know what it is. There is no name for it. You wonder if you're not going insane. You don't tell anyone out of fear of sounding paranoid. And if you do, you're not believed. There is nowhere to turn for information.

Then, by some means of communication, you're given a chance of serving it as a way of having it reduced or stopped. It's clear to me that this has been the experience of some of those now participating in this program. They were repeatedly assaulted by something they couldn't understand or explain, they were given an opportunity, and they took it.

In my opinion, this has nothing to do with lack of strength, morals, ethics, or intelligence. Some may have been attacked and recruited because they possessed those exact traits. So some of these people are victims themselves, regardless of where they are on the hierarchy of this program.

Voluntary Servitude

The rest of this section focuses on those who willingly choose to participate. In order for a well-meaning person to willingly participate in this program, they would have to be deceived. As previously mentioned, lies told by authorities play an important role in the recruitment process. But there are other forces at work such as group dynamics and human needs. There are some traits I've noticed which would predispose a person to this type of influence. First, a lack of understanding of authentic history. Secondly, a belief that they are well informed (naivety). Thirdly, (and probably most importantly) minimal pursuit of genuine personal development.

Psychologists, social workers, and personal growth instructors have concluded that there is a set of human needs that we all strive to fulfill. While experts may disagree on the number of needs, and the words used to describe them, they do agree that there is a core set of them. They help explain why people join groups in general, ranging from a local street gang to a yacht club. The words used to describe these needs are simply pointers to a feeling that people are looking to experience.

These needs transcend culture, age, and race. All human beings have these needs. So it's not a matter of whether they are met; they will be met, it is only a matter of how. I've included some of the basic human needs below. Words are sometimes a limitation, especially when explaining human needs. For that reason, I've combined several words to describe the same need. They are:

- Security/Certainty/Comfort/Safety/Confinement/Structure
- Uncertainty/Excitement/Adventure/Freedom/Variety
- Significance/Independence/Importance//Power
- Acceptance/Connection/Closeness/Belonging/Support
- Growth/Competence/Expansion
- Contribution/Service

Even though all of these needs must be met, there is a hierarchy of importance that each person chooses (usually unconsciously). A package which offers these needs will be appealing to most people. Citizen informants are obviously offered these needs in a compelling package. The Think Tanks which allegedly created this program are definitely skilled at exploiting these needs and manipulating large groups of people. They know *exactly* how to motivate people. Therefore, you can bet that their well-written propaganda material is filled with *power words* such as:

- Community
- Partnership
- Contribution
- Participation
- Service
- Responsibility
- Independence
- Growth
- Leadership
- Empowerment

There are the surface reasons we do the things we do, and there are the *real* reasons. The surface reason is a means to an end. The end reason is a state of being. I doubt if most of the people who participate in this program are aware of their true personal motives. In his book, *Vital Lies, Simple Truths*, Dr. Daniel Goleman wrote, "As one sociobiologist puts it, 'it is not difficult to be biologically selfish and still appear to be sincere if one is sufficiently ignorant of one's own motives.'"

In my experience, the primary motivating factors for citizen informants in Russia and Germany were *personal empowerment* and *contribution*. But probably more so, *personal empowerment*. For the most part, informants in Germany weren't paid, or were paid very little. Some were given gifts. But they believed that by participating they were *special*, they were *somebody*. They liked meeting with their handlers and being listened to. They did it for the deep human satisfaction of having, "one up" on the people they were watching. Despite their surface claims of patriotism, they were motivated by a *selfish* desire for *personal empowerment*. In essence, they received an *empowerment package*. It seems to me that the current situation is no different.

"USAonWatch through Neighborhood Watch empowers citizens in our communities with the opportunity to volunteer to work toward the safety of our homeland."
-George W. Bush

Sergei Kourdakov, a member of a Russian Citizens Brigade, stated that becoming a member was one of his *proudest moments*. He recalled how nice it was to be part of something he could *believe in.* Regarding the propaganda lectures which his group would regularly receive, he noted that the leaders "weren't sitting back, waiting for our enemy to destroy us from within." We were people "who admired action and we listened with admiration as we heard of the massive efforts being mounted." He exclaimed, "And to think that we were part of it!"

Referring to the East German IMs, Fulbrook wrote, "Many informants were committed and loyal citizens ... who were genuinely convinced that it was important to keep a watchful eye out for what they held to be subversive activities of dangerous 'enemies of the state.'" However, she adds, "[In] the majority of cases, the informant was persuaded into cooperating through a combination of feeling important, being attracted by small advantages, social rewards, material inducements or by a sense of adventure and stimulation to break an otherwise routine existence."

Speaking of adventure, it's clear to me that some of these people are having fun while stalking and harassing. It took me a while to categorize the facial expression that some of these people exhibit when I encounter them. Eventually I concluded that it was one of obsession—they're obsessed! Not necessarily with me, but with the act itself. In the book *Mobbing*, Dr. Carroll Brodsky, a psychiatrist, is quoted as saying, "Whether it occurs among nations or among individuals at the highest or lowest socioeconomic and political levels, harassment seems to be a social instinct ... [and] human beings fall easily into harassment behavior even when there seems no rational objective."

These citizens seem confident that a program sponsored by their Federal Government and endorsed by leaders in their local government must be moral. But lacking historical knowledge, they are unaware that they are playing a role in malicious process. When these groups sprung up in Germany and Russia, they were not created by a rogue element within the government they were created *by the government*, and run by local and federal law enforcement. All major components of the community were complicit. This means that trusted, and respected members of the community (who may have been operating on lies themselves), endorsed it. In other words, it was sold by the "good guys," who told these people they'd be protecting their country from the "bad guys."

From what I have noticed, most participants seem to require this type of structure, even though it is limiting, destructive and an obstruction to their development. This primitive, decaying thought pattern (*herd mentality, group think*), apparently left over from pre-historic times, has nothing to do with following orders from a superior, respect, participating in constructive group efforts, or working together for a goal. In reality, it is a subtle surrender of the individual's identity, and conformity to the *Group Mind.*

This type of surrender is very subtle and is usually unconscious. After it occurs, two things happen. First, each individual's thinking capacity is reduced to that of an adolescent. This has been called *regression.* Secondly, the limited product of the relinquishment merges itself into the group, which, itself functions as one entity called, the *Group Mind.* This *Group Mind* is ultimately limited and almost always insane.

"A hallmark of the person as group member, Freud saw, was the replacement of his self by a group self," stated Dr. Goleman. "The group archetype ... is the 'primal horde' a band of primitive 'sons' ruled by a strong 'father.' The particular schemas that constitute the group mind, in this view, would be those dictated by the father, a charismatic, strong leader." He continued, "The members of such a group, said Freud, relinquish their intellect to the chief... In the group mind 'the individual gives up his ... ideal [his will] and substitutes for it the group ideal as embodied in the leader.'"

In his book, *People of the Lie*, psychiatrist Dr. M. Scott Peck described groups as having a quality of "psychological immaturity." He wrote, "Individuals not only routinely regress in times of stress, they also regress in group settings. ... One aspect of this regression is the phenomenon of dependency on the leader. It is quite remarkable," he exclaimed. "It does not happen by a rational process of conscious election; it just happens naturally—spontaneously and unconsciously." Although most of the surrendered people are not aware of it, they have exchanged their *self* for what they perceive to be a more empowering *Group Mind.*

"Each individual possesses a conscience which to a greater or lesser degree serves to restrain the unimpeded flow of impulses destructive to others. But when he merges his person into an organizational structure, a new creature replaces autonomous man ... mindful only of the sanctions of authority."
-Dr. Stanley Milgram, 1974

Dr. Goleman observed, "Madness, said Nietzsche, is the exception in individuals, but the rule in groups. Freud agreed. In *Group Psychology and the Analysis of the Ego*, Freud wrote, 'A group is impulsive, changeable and irritable.' ... Freud saw people in groups as regressing to an infantile state as consequence of membership." He continued, "By 'group' Freud meant something that bordered on a mob." "Like individual evil, group evil is common," wrote Dr. Peck. "In fact," he warned, "it is more common—so common, indeed, it may be the norm."

After the individuals surrender themselves to the group, the entire herd merges into a single being. Dr. Peck explains, "This is because a group is an organism. It tends to function as a single entity. A group of individuals behave as a unit because of what is called group cohesiveness. There are profound forces at work within a group to keep its individual members together and in line." Although this type of surrender is an obstruction to personal growth, most of these people consider it a compelling trade for temporary artificial *empowerment*.(*)

empower
1. To invest with power
2. To equip or supply with an ability; [to] enable
-www.thefreedictionary.com

This "group cohesiveness" that holds groups together is described by Dr. Goleman as "mental homogeneity" which is made up of shared beliefs he calls "schemas." Quoting Freud, Dr. Goleman wrote, "What separates a 'group' from a random crowd is shared schemas," which he describes as "a common interest in an object [or person], a similar bias in some situation..." "The more in common a group shares ... the higher the degree of 'mental homogeneity' and therefore, 'the more striking are the manifestations of group mind.'"

So, if a group has been created for a specific purpose, and group cohesiveness is maintained by working toward a common goal, would the group create circumstances to foster its own cohesiveness in order to survive? The answer is yes. According to Dr. Peck, the most effective way of fostering group cohesiveness is to create an enemy. This helps to explain why dictators have usually created internal or external threats, as an excuse to form groups which would persecute their enemies. Again, this is always done *legally*, by the law enforcement of that time period. The effectiveness of this tactic is amplified when used within a *Specialized Group*, which was created to neutralize "enemies" in the first place.

Dr. Peck writes, "A practically universal form of group narcissism is what might be called 'enemy creation.'" He says, "Those who do not belong to the group (or club or clique) are despised as being inferior or evil or both." "If a group does not already have an enemy," he explains, "it will most likely create one in short order."

Enemy Creation

"It is almost common knowledge," maintains Dr. Peck "that the best way to cement group cohesiveness is to ferment the group's hatred of an external enemy. Deficiencies within the group can be easily and painlessly overlooked by focusing attention on the deficiencies of ... [others]." Dr. Goleman illustrates, "A group may implicitly demand of its members that they sacrifice the truth to preserve an illusion. Thus the stranger stands as a potential threat to the members of a group, even though he may threaten them only with the truth. For if that truth is of the sort that undermines shared illusions, then to speak it is to betray the group."

"The evidence for the collective defenses and shared illusions at work in groups ... is nowhere better stated than in Irving Janis's research on 'groupthink'" wrote Dr. Goleman. "The first victim of groupthink is critical thought. ... Facts that challenge the initial choice are brushed aside. ... Loyalty to the group requires that members not raise embarrassing questions, attack weak arguments, or counter softheaded thinking with hard facts. Only comfortable shared schemas [beliefs] are allowed full expression."

In her book, *Betrayal Trauma*, Dr. Jennifer J. Freyd wrote, "Indeed, becoming socialized to a culture may be partly dependent on one's falling into a kind of appropriate 'trance' in which one assimilates the norms and expectations of the members of the group..." She continues, "Such a state occurs in both small and large social organizations, in families and in nations, and the behavior may seem disturbing when viewed by an outsider ... as when 'normal' German citizens engaged in heinous acts under Hitler..."

We already understand why people join groups, and now we have an understanding of what happens after they do. But why do people, even if receiving an *empowerment package* surrender their *self* in exchange for being led (or misled)? Dr. Peck provides us with another answer, stating, "Most people would rather be followers. More than anything else, it is probably a matter of laziness." So, according to these professionals, this unconscious process works this way: *Because I'm a lazy coward, I'll give you my mind, and you give me some temporary empowerment.* As primitive and simple-minded as this underlying motivational factor may be, it seems to apply in most cases.

"USAonWatch empowers citizens to become active in homeland security efforts through participation in Neighborhood Watch groups."
-Citizen Corps program description

Birds of a Feather

People who operate on a particular frequency have a tendency to become aligned with acts and structures that emanate from that frequency. For instance, some of the largest historical atrocities that have taken place were not carried out by individuals, but groups, large groups. You can think of it as a negative energy field that continues to grow and use human beings as instruments to express itself. It almost always disguises itself as something good such as, "a great cause," and usually uses deception to enlist supporters.

The people who participated in those atrocities did not find their way into that structure and commit those acts by accident. Those people were attracted to that energy field because they operated on a frequency similar to it. The frequency of energy that I'm talking about here is not a high one; it is in fact primitive, but also popular. Some may call this destructive energy field, "evil." But I think "insanity" or "unconsciousness" are also accurate descriptions.

"Persons who have a weakened sense of self feel more secure in a group in which they find support," stated the book, *Mobbing*. "They often 'vulture,' lacking the courage to listen to their conscience. They find their identity by associating with someone believed stronger, respected, idolized." Dr. Westhues, professor of Sociology at the University of Waterloo, and author of the book, *Eliminating Professors* stated, "The only relevant reality at that point, far beyond any personality characteristics of the Mobbee ... is the pack, the herd, a number of individuals having surrendered their selves to something bigger."

McKinney described the Gang Stalking participants as "covert wannabees" carrying out what they think are covert activities, which are "perverted in the ultimate objective." And referring to the white-collar people who participate, Moret declared, "You would think it would not be so easy to co-opt them." My understanding is that most of these people seem to have a fragile internal support system, but rather than strengthening it by going through a process that is uncommon and somewhat frightening, they seek security outside of themselves. They thrive on recognition and acceptance, and will eagerly conform in order to meet these needs.

Obviously *keeping an eye* on people is a very big and empowering part of their lives. But if this vehicle were to be removed, they'd have a choice. One option is that they could go through a process of growth, which results in the development of power that is truly theirs and is with them under any circumstances. This approach is not taught in the mainstream. In fact, it is denounced! So the more common method is for them to quickly find one of the many readily available vehicles (groups), which will safely and easily offer them the temporary power they need.

I have not detected a category as far as age, occupation, religion, education, race, social standing, or financial status that these people can be placed in. But I have learned this: *They're the norm!* And despite their status, their minds are generally small and closed. I have no reservations about saying that the financial elite are using the small-minded, mentally ill, and deceived people to persecute the ones who are waking up or are already awake.

So, on the surface it would appear that their involvement is a result of carefully calculated decisions made by fully functional and informed adults, who participate out of patriotism and service to the greater good. However—upon further investigation, we find their involvement is born of a cowardly and selfish scapegoat mentality.

Dr. Peck referred to groups which are created for a specific purpose as *Specialized Groups*, and noted there is "subtle but definite scapegoating involved." He continues, "For the reality is that it is not only possible but easy and even natural for a large group to commit evil without emotional involvement..." "It happened in Nazi Germany. I am afraid it will happen again," he warned. It is happening again. The concentration camps are electronic; the physical intimidation has been replaced by more *refined* non-physical violence.

Some who willingly participate in this program may disagree with the analogy of Nazi Germany. They may say, "Oh, you're full of it, we're not torturing anyone." My reply would be that such a person is naive, in denial, or just lying. It's exactly the same evil, but more subtle. No visible torture chambers, but torture just the same. In fact it stems from the same financial interests that funded the Nazis—literally funded a holocaust... then lied about it. Anyone who has an open mind with no hidden agenda that has read my previous supporting evidence, probably knows that the people who created this are *monsters*.

Evil in Disguise

There is another category of people who willingly participate. They suffer from an acute type of insanity that could be described as *evil* or *psychopathic*. These people do it because they simply enjoy watching others suffer. They're not criminals. They pay their taxes on time, they pay their bills, and they go to church. They're members of your local community center. They're soccer moms and little league baseball dads. They're well groomed. They appear to be model citizens.

"They live down the street—on any street," says Dr. Peck. "They may be rich or poor, educated or uneducated. There is little that is dramatic about them. They are not designated criminals. More often than not they will be 'solid citizens'—Sunday school teachers, policemen, or bankers, and active in the PTA." He adds, "They dress well, go to work on time, pay their taxes, and outwardly seem to live lives that are above reproach." Dr. Peck reveals that these people can't be categorized by any existing "psychiatric pigeonholes," not because, "the evil are healthy," but "simply because we have not yet developed a definition for their disease."

He asks, "How can this be?" "How can they be evil and not designated as criminals? The key lies in the word 'designated.' They are criminals in that they commit 'crimes' against life and liveliness." "But" he says, "except in rare instances ... their 'crimes' are so subtle and covert that they cannot clearly be designated as crimes." It's usually not the size or illegality of their acts, but the consistency with which they commit them. He adds, "While usually subtle, their destructiveness is remarkably consistent." This pattern of consistent destruction, which Dr. Peck describes, seems to fit the people and organizations mentioned in Volume I perfectly.

Appearance is extremely important to these people. This is for concealment. They're petrified of someone finding out what they are. "Utterly dedicated to preserving their self-image of perfection, they are unceasingly engaged in the effort to maintain the appearance of moral purity," wrote Dr. Peck. "They worry about this a great deal. They are acutely sensitive to social norms and what others might think of them."

So according to Dr. Peck, it's as if evil people have something like a checklist of tasks, that they think "good" people do, and they go through the motions of performing these deeds so they can outwardly appear decent. They go through these motions not to be good, but to *pretend* that they are. "The words 'image,' 'appearance' and outwardly,'" states Dr. Peck, "are crucial to understanding the morality of the evil. While they seem to lack any motivation to be good, they intensely desire to appear good. Their 'goodness' is all on a level of pretense [fakeness]. ... It is, in effect, a lie."

Interestingly, one task that these *fake good people* usually engage in, as part of their cover, is destroying evil. "Strangely enough," stated Dr. Peck, "evil people are often destructive because they are attempting to destroy evil." "Instead of destroying others," however, "they should be destroying the sickness within themselves." He observed that they are often "busily engaged in hating and destroying ... usually in the name of righteousness." Therefore, the Hidden Evil, with its Neighborhood Watch groups and silent electromagnetic torture, is the perfect expression for their sickness. The liars—the fakes—can harm while appearing to help. They can hide within it, because the program itself—is a lie.

Ordinary Evil

In my estimation, most of the citizens now participating in Gang Stalking are not evil. But you don't need to be evil to participate in evil. "Even civilians will commit evil with remarkable ease under obedience," remarked Dr. Peck. "As David Myers described in his excellent article *A Psychology of Evil* (*The Other Side* [April 1982], p. 29): 'The clearest example is Stanley Milgram's obedience experiments. Faced with an imposing, close-at-hand commander, sixty-five percent of his adult subjects ... would deliver what appeared to be traumatizing electric shocks to a screaming innocent victim in an adjacent room. These were regular people—a mix of blue-collar, white-collar and professional men.'"

Dr. Stanley Milgram's obedience experiments, were originally conducted at Yale University during the 60s, and then conducted in other parts of the world. The tests were done to see how many ordinary people would commit acts of violence against an innocent person. The tests consisted of a teacher and a learner scenario. The teachers were told that they would be administering a test to a student or learner, and if the student answered a question wrong, they were to shock him. In reality, the learner was an actor, and the real subject of the experiment was the teacher. The true nature of the experiment was to see how much pain an ordinary person would inflict on an innocent victim with a heart condition, under the direction of an authority figure.

In the December 1973 issue of *Harpers Magazine*, under an article entitled, *The Perils of Obedience*, Dr. Milgram wrote, "The point of the experiment is to see how far a person will proceed in a concrete and measurable situation in which he is ordered to inflict increasing pain on a protesting victim." The only way for the subject to disobey and halt the experiment was for them to make a clear brake with authority.

Before the experiments Dr. Milgram asked a variety of people including, psychiatrists and faculty in the behavioral sciences, what the outcome would be. They answered that only a "pathological fringe of about one in a thousand would administer the highest shock on the board." "These predictions," stated Dr. Milgram "were unequivocally wrong." 65% of the adult test subjects administered the full 450 volts to the victim.

One colleague dismissed the experiments, claiming that once "ordinary people" outside of the Yale campus were selected, the numbers would decrease. So, the experiments were conducted by other scientists in places such as Princeton, Munich, Rome, South Africa, and Australia. The level of obedience increased to has high as 85%.

Dr. Milgram warns, "The most fundamental lesson of our study," is that "ordinary people, simply doing their jobs, and without any particular hostility on their part, can become agents in a terrible destructive process." But, even if they're aware of the destructive process they'll still participate under the right circumstances. He adds, "Even when the destructive effects of their work become patently clear, and they are asked to carry out actions incompatible with fundamental standards of morality, relatively few people have the resources needed to resist authority."

What else did they discover? Well, fortunately many protested before administering the shocks, but unfortunately even though they were "totally convinced of the wrongness of their actions," they "could not bring themselves to make an open break with authority," wrote Dr. Milgram. These psychiatrists also found that most people didn't like inflicting pain on innocent people, but did enjoy the satisfaction of pleasing the authority figure. Contrast this to citizen informants looking for a pat on the back by local authorities for their *service*. If this type of motivation sounds small-minded, remember, it's what motivated adults in other countries to do the same. They were, "proud of doing a good job, [and] obeying the experimenter under difficult circumstances," proclaimed Dr. Milgram.

Who was this authority figure? Well, in the experiments he was portrayed as a doctor with a stern voice, telling the subjects that they must continue. Milgram noted that "he had almost none of the tools used in ordinary command structures." In other words, this authority figure did not portray himself as a military officer, an FBI or NSA agent, or a local official. "For example," says Milgram "the experimenter did not threaten the subjects with punishment," such as "loss of income, community ostracism, or jail," but despite these limitations, "he still managed to command a dismaying degree of obedience."

Variations of the experiment were carried out and levels of obedience increased when the responsibility was diffused. The act of administering the shock was spread to multiple people. "Predictably," says Milgram "they excused their behavior by saying that the responsibility belonged to the man who actually pulled the switch." The person who pulls the switch is simply following orders.

He continues, "This may illustrate a dangerously typical arrangement in a complex society: it is easy to ignore responsibility when one is only an intermediate link in a chain of actions." Similarly, although many of these citizen informants may not *pull the switch*, they're an essential part in a relay that results in torture and murder. Dr. Milgram adds, "Perhaps this is the most common characteristic of socially organized evil in modern society."

Yale Alumni Magazine produced an article on Dr. Milgram's findings. Written by Dr. Phillip Zimbardo, it appeared in the January/February 2007 issue and was entitled, *When Good People Do Evil*. Dr. Zimbardo asked, "Obviously, you would have dissented, then disobeyed and just walked out. You would never sell out your morality," he asked, "Right?"

"Milgram," continued Zimbardo, "once described his shock experiment to a group of 40 psychiatrists and asked them to estimate the percentage of American citizens who would go to each of the 30 levels in the experiment. On average, they predicted that less than 1 percent would go all the way to the end, that only sadists would engage in such sadistic behavior... They could not have been more wrong. ... The vast majority of people shocked the victim over and over again despite his increasingly desperate pleas to stop."

Frighteningly, when the experiments were altered, "Milgram was able to demonstrate that compliance rates could soar to over 90 percent," proclaimed Dr. Zimbardo. "In one study, most college students administered shocks to whimpering puppies when required to do so by a professor." He added, "His classic study has been replicated and extended by many other researchers in many countries." When a more pronounce authority figure is used, compliance increases. It also increases when the act of inflicting pain is fragmented among many people. And the greater the distance, the easier it is to inflict pain.

Dr. Zimbardo wrote of another study where a group of twenty high school students joined a history teacher's authoritarian political movement, and within a week had expelled their fellows from class and recruited almost 200 others from around the school. The article continued, "Motives and needs that ordinarily serve us well can lead us astray when they are aroused, amplified, or manipulated by situational forces that we fail to recognize... This is why evil is so pervasive."

"Nothing is more terrible than to see ignorance in action."
-Goethe

These experiments clearly show a significant flaw in the logic of the masses which allows persecution, genocide, and holocausts to occur. In fact, Dr. Milgram originally conducted the experiments to see how an entire country could go along with persecution and mass murder in Nazi Germany. This is not a problem with the "evil elite" who want to control us, but a problem with humanity. This is a problem with society.

In a March/April 2002 article entitled, T*he Man Who Shocked The World*, *Psychology Today Magazine* described Dr. Milgram as, "the man who uncovered some disturbing truths about human nature," by conducting "one of the most significant—and controversial—psychological studies of the 20th century."

These studies, they stated "demonstrated with jarring clarity that ordinary individuals could be induced to act destructively," and "need not be innately evil or aberrant to act in ways that are reprehensible and inhumane." The article concluded, "Milgram's obedience experiments teach us that in a concrete situation with powerful social constraints, our moral sense can easily be trampled."

Milgram's researched led him to conclude that there is a protocol used by leaders of countries that artificially creates an environment of situational power, which will foster the participation of the public in carrying out acts of evil. It is basically a protocol for mass persecution. Speaking of this, Dr. Zimbardo noted, "Milgram crafted his research paradigm to find out what strategies can seduce ordinary citizens to engage in apparently harmful behavior. Many of these methods have parallels to compliance strategies used by 'influence professionals' in real-world settings, such as salespeople, cult and military recruiters..."

In other words, these methods are a type of mass *mind-control.* Once more, in the past, the masses did not detect that they were being negatively influenced to commit these acts, just as they're not aware of it now. As Professor Marrs described, one key aspect of a dictatorship, is that the target population may be unaware of its existence. The protocol includes:

(A) "Prearranging some form of contractual obligation, verbal or written, to control the individual's behavior in pseudo-legal fashion." In Germany they signed wavers when recruited into the Ministry for State Security, in Russia they were sworn to secrecy. Obviously the current informants go through a similar process.

(B) "Giving participants meaningful roles to play ... that carry with them previously learned positive values..." In East Germany the IMs were recruited for "state security." In Russia they were "Citizen's Brigades," maintaining civil order.

(C) Using semantic deception to alter "the frame so that the real picture is disguised" from "hurting victims" to "helping." Clearly the current participants are told that they're *helping* their communities and country, not *helping to murder* innocent people. Again, similar propaganda was used in Germany and Russia.

(D) "Creating opportunities for the diffusion of responsibility or abdication of responsibility for negative outcomes..." There are two considerations. First, the fragmenting of responsibility increases group complicity and removes accountability, which results in people going above and beyond what they'd normally do as individuals. In this situation, no one is directly responsible for the murder of a person. Secondly, when an authority figure takes responsibility for the acts by making them OK, accepted, or legal, then complicity increases.

(E) Possibly the most important part of this protocol is the offering of the "big lie," which Dr. Zimbardo says is used to "justify the use of any means to achieve the seemingly desirable, essential goal."(**) "In social psychology experiments, this is known as the 'cover story,'" says Dr. Zimbardo. This lie, he declares, "is a cover-up for the procedures that follow, which do not make sense on their own." He describes the real world equivalent as "an ideology," such as a nation relying on "threats to national security," before suppressing political opposition.

"An evil exists that threatens every man, woman, and child of this great nation. We must take steps to ensure our domestic security and protect our homeland."
-Adolph Hitler

An article in the July/August 2004 issue of the *American Scientist* entitled, *Milgram's Progress*, stated, "In a television interview in 1979, Milgram said that he eventually came to the conclusion that, "If a system of death camps were set up in the United States of the sort we had seen in Nazi Germany, one would be able to find sufficient personnel [to operate] ... those camps in any medium-sized American town."

But we've come a long way since the 60s, we don't think like that anymore, right? Besides, those types of things happen in *other* countries... Unfortunately this is false. People have not changed in thousands of years. Technology has—not people. Again, the same financial interests that funded those dictatorships are allegedly responsible for the current situation.

In fact, years later *ABC News* re-created these obedience experiments in its own study and in January 2007 published their findings in an article entitled, *The Science of Evil*. They wrote, "Primetime wanted to know if ordinary people today would still follow orders, even if they believed their actions were causing someone else pain. Would as many follow the seemingly dangerous and painful orders as in the original experiment?" They found that obedience held at 65%, and in some cases increased to a surprising 73%. Like the original studies, theirs subjects were composed of an ethnically diverse group with an unusually high level of education.

Referring to specialized groups, Dr. Peck wrote, that the "group will remain inevitably potentially conscienceless and evil until such time as each and every individual holds himself or herself directly responsible for the behavior of the whole group. ... We [the general mass] have not yet begun to arrive at that point."

"With numbing regularity good people were seen to knuckle under the demands of authority and perform actions that were callous and severe. ... A substantial proportion of people do what they are told to do, irrespective of the content of the act and without limitations of conscience, so long as they perceive that the command comes from a legitimate authority."
-Dr. Stanley Milgram, 1965

Parasitic Behavior

There is another motivational factor worth mentioning. Possibly, the greatest factor. During these covert group attacks, there is an energy transfer that these individuals, whether they know it or not, are looking for. The systematic vulturing during Mobbing and Gang Stalking campaigns is an intended robbery of a person's life-energy. This is no different than sticking a needle in someone's arm and stealing their blood. Eastern philosophy refers to these people as *Sappers*.

A *Sapper* is a person who is too sick, weak, or underdeveloped to create their own life-energy. So they continually find themselves in circumstances where they steal or *sap* energy from a being that has already assimilated it into a usable form. This is similar to an infant eating pre-processed food. Society is full of them. They are completely unconscious of this tendency. I also refer to them as *Psychic Fleas*.

Summary

When we examine the underlying factors which motivated citizen informant groups in the past, we find that feelings of personal empowerment, and a sense of adventure, figured largely. Present-day citizen informants, in all likelihood, are motivated by these factors. While having their personal human needs fulfilled, they may not recognize that they're part of a destructive process. Most ordinary people will commit evil regularly under the right conditions.

People with weak support structures feel more secure in a group. The group offers protection in exchange for conformity and obedience. Groups created for a specific purpose engage in scapegoating, which includes the creation of enemies to foster its existence. The group functions as a single entity with the psychological capacity of an adolescent. There exists an actual recipe for gaining mass complicity in persecution, where this entity can be used.

Endnotes

* It's almost as if they're enabling the disabled in order to disable the enabled.

** Coincidently, one alternate name I contemplated for the Hidden Evil was, "The Big Lie."

Chapter 22

Potential Targets

As I've already demonstrated, the people controlling the United States and other developed countries are not good people. Once you understand how these controlling elite operate, it's easy to see why they are deliberately targeting innocent people. On September 12, 2006, the *Associated Press* published, *Official Touts Nonlethal Weapons for Use*, where they announced, "Air Force official says nonlethal weapons should be used on people in crowd-control situations."

They continued, "Nonlethal weapons such as high-power microwave devices should be used on American citizens ... the Air Force secretary said Tuesday. ... Domestic use would make it easier to avoid questions in the international community over any possible safety concerns, said Secretary Michael Wynne." Wynne stated, "If we're not willing to use it here against our fellow citizens, then we should not be willing to use it in a wartime situation."

In the June/July 1987 issue of *Peace Magazine*, under an article entitled, *Zapping the Movement*, Congressman James Scheur is quoted as saying, "We are developing devices and products capable of controlling violent individuals, and entire mobs without injury. ... We can tranquilize, impede, immobilize, harass, shock, upset, chill, temporarily blind, deafen, or just plain scare the wits out of anyone the police have a proper need to control and restrain."(*) The military and a member of congress have now openly declared that these weapons should be used on civilians.

The Sun Journal ran an article on September 28, 1992, entitled, *Woman Fears Government Zapping*, where they described a 72 year-old woman that had been harassed in multiple states and countries, allegedly by the FBI and CIA. She visited various doctors in Richmond, VA, who told her that nothing was wrong. The harassment continued when she visited doctors in England, Sweden, Finland, and Canada.

"Electronic government weapons are zapping burns in her face ... and emit rays that stab her back and make her brain feel swollen," they wrote. "She says she gets zapped in her home, her car and even in church on Sunday mornings. ... She does not know why she is a target." They continued, "[She] says she is sure the powers that are tormenting her are evil ... [and doesn't] know why they are after her."

"Our government now is controlled," she said, and added, "I think the CIA, the FBI and the military are evil."

If you're targeted, no one will tell you so. Some people conclude they were harassed for years before they realized what was happening. Multiple people within a family can be targeted, which indicates a pattern suggesting that these harassment campaigns are intergenerational. Many people don't know why they're targeted or exactly when it began. "No TI should look for a reason why this is going on," warned McKinney. "It's a serious, serious mistake."

As a result of surveillance, the people overseeing these harassment squads know this is nothing about monitoring a threat to national security, and everything to do with the deliberate misuse of resources for no good reason. When you see the words, "threat to national security," think "threat to corporate control." McKinney's analysis leads her to conclude that most of those targeted are not and never have been a threat. Moret describes this program as an "absolutely terrible secret policy," and warns, "Americans need to understand that they are the primary targets of this war on terror."

This is the elite's way of removing people from society that they dislike or see as a nuisance. Once you research this, and find out what they've done, how they've repeatedly lied, and what their intentions are, then this accusation doesn't seem outlandish, but makes perfect sense. McKinney hinted that the reasons of national security blanket might be the justification for these harassment campaigns, and spoke of the "absence of clear definitions" of terms such as "national security" and "national security risk." Similarly, in East Germany and Russia, people were labeled threats to *state security*, *enemies of the state*, and *enemies of the people* to justify persecution.

"In addition to criminal and terrorist organizations," wrote Dr. Alexander, "other groups may emerge that have the ability to threaten national and regional security." He lists some of these groups as fundamental religious groups, groups looking for homelands, starving masses, those who are economically deprived, and those who believe they are socially oppressed. "These threats are both internal and external," he says.

In his book, *The Franklin Cover-up*, Senator DeCamp wrote, "The government's legitimate concern with national security has been turned into a banner under which government officials and judges and agencies and politicians can, and do get away with almost anything and everything."(**)

I knew law enforcement and the intelligence community had their problems, like most institutions do. I had also heard scattered stories of people being harassed by government agencies but I attributed it to their own fault. I figured they must have been some serious troublemakers in order to provoke it. I thought the occasional media leaks were pretty much all there was to it. I trusted that the "watch-dog" media kept those agencies in line and were perhaps even a little intrusive (preventing these agencies from doing their jobs).

I believed that they would never spend resources to target innocent people, especially a *nobody*. I thought that they had better things to do than to waste resources targeting innocent citizens. I was wrong. I never realized that those media leaks were just the tip of the iceberg or how big and organized this corruption was.

There is evidently a list that people are added to. Once someone is added to this list, they are open to all manner of experimentation with surveillance and Directed Energy Weapons. Once someone is labeled a *threat to national security*—anything goes. There are no laws preventing this. No mainstream organization to this date has made a genuine effort to expose this program. It appears that there are multiple ways a person can be added to this list.

Whistleblowers

Government whistleblowers or whistleblowers of a corporation that is closely associated with the government may be targeted. Some people are convinced that they were targeted because they exposed corporate or government corruption. Leuren Moret, for instance, was targeted after exposing hazards relating to depleted uranium at the Lawrence Livermore National Laboratory.

Constantine described, "The whistle-blower is forced out, frightening workmates and supporters. ... The assault continues" he says, "until the target is left discredited, exhausted, in poor health, financially crippled, his career in ruins." He explained that the objective is to portray them as "incompetent, disloyal, troublesome, mentally unbalanced or ill."

Protestors

Civil rights activists have a history of being targeted by similar programs. "The records show that the vast majority of the targets of domestic covert action have engaged only in peaceful protest," declared Glick in *War At Home*. "They do no harm to anyone's health or safely.

The only danger they pose is to the status quo. Their only weapon is the power of their words and the threat of their good example."

On March 10, 1986, *The Guardian* stated in an article called, *Peace Women Fear Electronic Zapping at Base*, that about 40 people experienced headaches, dizziness, an inability to concentrate, and memory loss after protesting outside a U.S. military base in England. "Doctors are compiling a report on the condition of a number of Greenham Common peace women who have had symptoms which are consistent with the known neurophysiological effects of electromagnetic waves, or low level radiation," stated the Guardian.

The article continued, "Dr. Stephen Farrow, Chairman of the Medical Campaign against Nuclear Weapons said yesterday: 'We are now compiling evidence about the claims made by the women.' Dr. Farrow, who is senior lecturer in epidemiology at the University College of Wales Medical College, said that academic research into similar claims was being conducted in Canada."

An article in the *Associated Press* on April 15, 1995, entitled, *Government Wants Me Dead*, reported that a lawyer named Linda Thompson filed a complaint at the Marion County Prosecutors Office stating that the government was trying to kill her with radio frequency weapons. The article went on to explain how Thompson claimed that six of her friends had already been killed this way.

According to a December 4, 2003 article in *The Village Voice*, entitled, *J. Edgar Hoover Back at the 'New' FBI*, even peaceful protesters are now being labeled as potential "terrorists." One need only look at the objectives of the Think Tanks which clearly state that they intend to abolish the constitution and merge America into a single world government. Therefore, anyone who publicly opposes the New World Order, or any of its related agendas such as the WTO, NAFTA, CAFTA, FTAA, the Federal Reserve, etc., may be targeted. Anyone involved in the 9/11-truth movement is a potential target.

Independent investigative journalist Greg Szymanski, who has written for the *American Free Press*, and has appeared on the *Republic Broadcasting Network* and the *Genesis Communications Network*, interviewed a man that has been targeted due to uncovering some information regarding 9/11. Apparently this man has been stalked by helicopters, hit with microwave weapons, and has been on the receiving end of death threats and multiple attempted murders, including staged accidents.

The Seattle Times ran an article in April of 2000, called, *Crowd-Control Cookery*, which described, when "unruly demonstrators disrupting World Trade Organization talks clashed with police," they "had at their disposal a new ... arsenal designed to sting, stun, entrap, immobilize, sicken, knock the wind out of—but not kill—the assailants, suspects, agitators or enemies they are used against." They continued, "They're referred to collectively as non-lethal weapons, and police and military units are increasingly using them as they try to limit the use of deadly force. ... Many of the nation's major urban police forces already use some of these weapons..."

This program may be particularly useful for subduing people who are too high profile, or for whatever reason can't be outright murdered. Say, for instance, if someone is quite troublesome to the elite, they can be given cancer, and psychologically tortured to impede their progress. Or in the words of Dr. Alexander, the program can be used to "neuter" them. Parents and children of targeted individuals are also being targeted for apparent purposes of intimidation. All done under the banner of "national security."

The Cleveland Plain Dealer ran an article on in June 1991, entitled, *Psychiatrist Testifies at Mom's Hearing*. Apparently a woman in Ohio killed one of her daughters to spare her from further trauma, after they had been targeted for two years with torture, consisting of burns and sexual degradation by V2K. Prior to this she wrote a multitude of letters to the pentagon, government officials, and talk show hosts describing her plight.

Her doctor told the Dealer, "she has a number of delusions ... [about] being repeatedly attacked and tortured by government agencies." He continued, "[her] unwillingness to understand her illness kept her from weighing the possibilities of her defense." Her attacks reportedly continued while she was imprisoned.

McKinney wrote of one protester who received a call stating that, "if she valued the lives of her children," then she would drop her public opposition to a company's installation of high power lines. "Since receiving that threat," declared McKinney, "the individual's 11-year old daughter has been reduced to extremes of pain, resulting in her recurrent hospitalization for treatment of illnesses which cannot be diagnosed. It is now also apparent to this individual that her three-year-old son is on the receiving end of externally-induced auditory input [V2K]."

Similarly, the Stasi would target families. Fulbrook described, "The children and relatives of dissidents might be subjected to harassment and personal disadvantage." She added, "People could emerge with their health, self-confidence and future working lives damaged beyond recovery."

Experimentation

Part of the Hidden Evil seems to be testing the latest in surveillance and Directed Energy Weapons. MKULTRA, a known mind-control project, with 149 sub-projects, was conducted in universities, hospitals, military institutions, and prisons. According to author Jim Keith, under MKULTRA about 23,000 people were traumatized, but researchers in this are see this as an extremely conservative number. Since many of the records have been destroyed we will probably never know. Likewise, the Stasi burnt out many paper-shredding machines destroying their victim file evidence.

The paper trail for these experiments ended around 1984. But evidence reveals that it never stopped. "It would appear that the CIA's and FBI's Operations MKULTRA, MHCHAOS and COINTELPRO," were "merely driven underground," stated McKinney. Labs are no longer necessary as Directed Energy Weapons can be used at a distance, and plain-clothed citizen brigades are complicit in public harassment.

Recruitment

It appears that some people are targeted with the intent to recruit them as informants after they've been broken. Both the East German Stasi and the CIA would use constant harassment campaigns on people they wanted to recruit as informants, or people knew they couldn't recruit and wanted to destroy.

Political Movement

Although whistleblowers, activists and others have been targeted, the scope of this worldwide program is too big for it to be limited to a parallel justice system. As previously illustrated, there is probably a political agenda attached to the Hidden Evil. According to both McKinney and Moret, it has moved beyond experimentation and is being widely used on the civilian population in all NATO countries. Dr. Kilde adds, "Today ANYONE can become a target, even those who invented the system." [Emphasis in original]

The controlling-elite are not individually picking targeted individuals, so there must be people on the local level, charged with the task of identifying people to target. McKinney has found that Corporate America is another medium for detecting TIs. This is not surprising because many of them run surveillance operations for government agencies.

McKinney declared that in the past 15 years, she has "had occasion to observe many, many, many instances of individuals in the corporate environment being singled out and targeted." It seems there are people in communities and businesses who identify people to be put under surveillance. This is very similar to what unfolded in East Germany, where workplaces, schools, and hospitals were complicit in surveillance operations and filled with informants.

What are the qualities that local agents are to look for when profiling targets? We may never know for sure. But Annie Earle's non-profit organization, An Ethical Light, has done some research on those alleging to be targeted. She acknowledged, "There does seem to be a certain profile among the victims we have studied." She describes them as "generally very highly intelligent people..." She says that the people her organization deals with "represent a wide range [of] every type of profession you can imagine."

The women who are singled out tend to be independent, intelligent, and confident professionals. "There's a heavy predominance of those types of woman in the TI community," affirmed McKinney. Men are in a smaller proportion, and tend to be non-conformists. They usually have a "sense of self-esteem and pride that seems to *invite* targeting," she says. Remember, the MfS was a criminal federal police force that deliberately singled out people exactly like this on a massive scale. Others that have been targeted include:

- Those who have had a bad breakup with an ex-spouse who has influence
- Criminals (Targeting known offenders may strengthen the illusion that the program is legitimate)
- Those who have reported a crime where the perpetrator was a member of an organized crime ring
- Landowners targeted by government or industry for land seizures
- Gays, minorities
- Inventors awaiting a large payoff
- Mentally disabled
- Those who have had arguments with neighbors who have connections

- Those awaiting a large insurance claim
- Convenient targets of opportunity
- Those with special talents or abilities
- Group members who act in unacceptable ways
- Independent freethinkers who are not part of the herd
- Those who are perceived as vulnerable or weak
- Those opposing any element the New World Order

The pattern that is unfolding indicates that many targets are people who tend to be emotionally developed, self-confident, independent, free thinkers, artistic, people who don't need the approval of others, and those not prone to corruption. They're people who don't need to be part of a group to feel secure—generally people who are not part of the herd. The financial elite throughout history have consistently *hated* and *feared* these individuals. There's one other type of person that the financial elite hate—truth tellers. They absolutely *hate* them. This is apparently why whistleblowers are targeted.

Those on the local level, who are given the authority to identify people to be targeted, are evidently a reflection of their military-industrial handlers. The profile of those who are likely responsible for the Hidden Evil, has been established as: greedy, deceitful, psychopathic, and cowardly. Therefore, envy, jealousy, and fear may also play a significant role in the selection of targeted people. On a smaller but similar scale, Mobbing targets are selected for these exact reasons. The book *Mobbing* states, "Envy, jealousy, aspirations, and being challenged are reasons individuals mob." They charge, "[people] might resent someone for performing better, for looking better, for being more liked. They fear others' competence."

Eugenics

For some reason, there are considerably more women targeted than men. Why would women be targeted at a much higher percentage than men? This is another question we may never have the answer for. So this is partially speculation. But one possible answer is that the elite have sponsored eugenics projects worldwide for decades. Removing fertile females from a target population is apparently a standard eugenics procedure. Families such as DuPont, Harriman, and Rockefeller have funded projects for population control.

"It was John D. Rockefeller III who was appointed by Richard Nixon as chairman of the newly created Commission on Population

Growth and the American Future," stated Allen. He quotes Rockefeller as saying, "Rather than think of population control as a negative thing, we should see that it can be enriching." Allen contends, "Curbing population growth is just part of the Rockefeller war on the American family."

Author Marrs wrote, "The Harrimans ... along with the Rockefellers funded more than $11 million to create a eugenics research laboratory at Cold Spring Harbor, New York, as well as eugenics studies at Harvard, Columbia, and Cornell." And according to author Jim Keith, eugenics programs in Nazi Germany were organized and funded by the Rockefeller Foundation.

The Hidden Evil is probably not employed strictly as a means of population control due to its delayed results and the amount of resources it consumes. However, because it is life-destroying in nature, the financial elite may see the targeting of women as an accent to their existing eugenics programs. Apparently, these eugenics projects are ongoing. Author Marrs declared, "Eugenics work, under more politically correct names, continues right up today." Keith agreed, writing, "In 1905, in the United States, the Rockefellers and Carnegies constructed the Eugenics Records Office at Cold Springs Harbor, New York, where genetic research ... is still being done."

They're Monsters

The last reason may be very difficult for some people to accept because it is simple and sadistic. It is this: They are picking people out of communities and violently murdering them because that's the type of people they are. This mind-set is probably completely alien to most people. Because most of us don't entertain these thoughts, it's difficult to imagine how another person could. But, insanity *is* the reason. Evil *is* the reason. This will be easier to understand when we explore the Satanic and psychopathic factors in future chapters.

Summary

Genuinely decent people are being systematically targeted by this cabal which, through propaganda and deception, has extended its influence all the way down into the neighborhood level. If you're a protestor, whistleblower, or they just don't like you, you can be labeled a "threat to national security." Once that occurs, you may have to watch as your children are slowly tortured and murdered with this technology. You will also be Gang Stalked by citizens' watch groups composed of men, woman, and children. The military, a member of congress, and the alleged creators of these weapons have publicly said that they will be used on civilians.

Endnotes

* When Congressman Scheur uses the word "we" I tend to believe he's referring to himself and other members of congress, or himself and the military/police, or both.

** Shockingly, during his investigation of an elite-sponsored pedophile ring with ties to Washington D.C., DeCamp discovered that the children were told that their participation was in aid of "national security."

Chapter 23

Effects

McKinney stated that the objective is to "force the individual to commit an act of violence, whether suicide or murder, under conditions which can be plausibly denied by the government." The antagonistic public Mobbing of a target is apparently intended to provoke an outburst, so they may end up incarcerated or institutionalized. Moret's analysis is similar. She says, "The purpose is to completely isolate and remove all social support, economic support, and if possible, to drive these targets to suicide." "Isolation of the individual from members of his/her immediate family ... [is] virtually assured," proclaimed McKinney, "when highly focused forms of electronic harassment commence."

Other objectives appear to be to separate a person from friends and family, keep them unemployed, induce homelessness, and reduce the quality of life so much that they suffer a nervous breakdown, end up medicated, or hospitalized. McKinney added that other "long-term objectives" of these harassment campaigns, appears to be to "induce a sense of perverted 'loyalty' toward the very agencies engaged in the individual's harassment..."

Moving to a completely new location will not stop this. All normal escape mechanisms have been removed in all of the NATO countries. Reportedly, even when hospitalized, targets are still harassed. Referring to the futility of moving to another location, McKinney said "Your protocol follows you wherever you go so it's a waste of time." "All of these harassment techniques are recurrent, they're non-sequential, and they are overlapping," proclaimed Moret. "And basically it's forever, they never leave you alone again. You're even followed out of the country, [so] it happens in other countries."

Interestingly, if someone under investigation killed himself, most within the sphere of the person's life would probably think this happened because they had something to hide, not because they were repeatedly injured. Therefore, bogus investigations may also provide a cover for an induced suicide by creating the appearance of a "guilty conscience." "Suicides might also qualify as 'staged accidents,' particularly where 'plausibly deniable' government involvement has been surfaced," wrote McKinney.

Symptoms

Because the group harassment for Mobbing and The Hidden Evil are the same, so are the symptoms. Some of the following symptoms have been taken from the book, *Mobbing*. They include:

- Nervous breakdowns
- Severe depression
- Severe panic attacks
- Heart attacks
- Other severe illnesses
- Suicide
- Violence directed at third parties
- Going postal
- Uncontrolled acting out
- Feeling suicidal and/or homicidal
- Persistent anxiety
- Fatalistic outlook on life
- Frequent or longer sick leave or disability at work
- Isolation

According to Moret, Munzert, and McKinney, other reported symptoms and situations are:

- Forced onto medication for anxiety/depression
- Imprisonment
- Forced into mental hospitals
- Homelessness
- Framings
- Staged accidents
- Death

When I worked in Boston for about five hears, each morning I'd pass by homeless people on the corner asking for money. Usually I'd be thinking to myself, "Now there's a perfectly healthy, able person. Why don't they just get a job? I mean they're not lazy because they're up at 7:00AM every day, standing on this corner asking for money, why not be doing something constructive while earning money?"

It had never occurred to me that some of them were probably Blacklisted and repeatedly Mobbed out of the workforce. I never knew such a mechanism existed. Most targets would be homeless if it weren't for friends or relatives. So when you see someone holding a sign that reads, "Will work for food," know that they may be targeted by The Hidden Evil.

It may be difficult for a person to conclude that they've been targeted, and these groups may gradually turn up the intensity of their harassment over a period of time. This is the equivalent putting someone in a pot of water, and slowly heating it until the water is boiling, i.e. they need medication for anxiety or depression. "Many, many, many thousands, no doubt, are involved [targeted]," declared McKinney. "But I would say the bulk of those are running to their doctors and taking totally unnecessary prescription drugs to cure ailments that in fact don't exist."

Earle noted that targeted people "are experiencing extreme traumatic stress." The targets she said, "are subjected to a level of stress that the average American cannot even imagine. They are constantly bombarded with auditory information, with crises that have been created." Dr. Munzert declared, "The victims experience extreme, unbelievable things; almost no one can believe their reports." Once more, the covert psychological torture that targets experience is identical to what East German targets received.

Former Stasi Chief Markus Wolf announced that the "psychological effects" of the Stasi's harassment campaigns, "were probably more damaging, ultimately, than any physical torture would have been." *The Bulletin of the Atomic Scientists* stated on July 1, 1999, that, "For many, the discovery that their best friends or their own family members were betraying them to the Stasi was worse, in its own way, than torture in an internment camp."

So far we've focused on illnesses that arise out of psychological harassment. But in addition to this, targets often experience physical symptoms resulting from Non-lethal attacks. McKinney described that many targets experience, "Rapidly deteriorating health, generally of a digestive nature," with "severe gastrointestinal disturbances."

One common DEW symptom is thermal heating which leaves a hot spot on the skull. "The objective of using microwaves," says McKinney, "would be to inflict extremes of pain, [and] to cause thermal heating." "Usually the victims experience pain ... [in the] brain, heart, [and] stomach," says Dr. Munzert. "It's a little bit like being heated in the interior of the body."

"And on the skin" he adds, "victims experience [what feels like] hundreds or thousands of stings or pins." If the energy is intense, "the inner organs begin to vibrate," he says. Dr. Munzert also states that these attacks are masked because these symptoms and illnesses occur naturally. This masking effect is compounded by the fact that most people are *not aware* that these weapons exist or of their capabilities. Some other DEW symptoms include:

- Memory loss
- Dizziness
- Headaches
- Extreme fatigue
- Sleep deprivation and disruption
- Jolts and jerks to muscles (the equivalent of being shocked)
- Abdominal pain/nausea
- Mental confusion/inability to concentrate
- Stings
- Burning sensations
- Heart attacks
- Strokes
- Aneurysms
- Cancer

If you see a mental health professional as a result of this covert violence, you may be misdiagnosed with a mental disorder. People with mental disorders are stereotypically seen as incompetent and violent, according to a report entitled, *Stigma of Mental Illness Still Exists*, by the Center for the Advancement of Health. This will probably follow you for the rest of your life.

"Frankly I strongly recommend that you keep your faculties together and avoid going to see physiatrists and psychologists, because the pattern that is evolving is that they are highly complicit in these operations," advised McKinney. Dr. Munzert agreed, indicating that care must be taken when explaining this to doctors, in order to avoid ending up in the "lunatic asylum." Reportedly, the harassment even continues inside hospitals and prisons.

According to The Wheeler Clinic's article, *Stigma and Mental Illness*, if you have been diagnosed with a mental disorder, you may have to answer to this on job/school applications, or for insurance. The article stated, "Stigma keeps people from getting good jobs and advancing in the workplace. Some employers are reluctant to hire people who have mental illnesses. Thanks to the American with Disabilities Act, such discrimination is illegal."

"But" they add, "it still happens! Stigma results in prejudice and discrimination. Many individuals try to prevent people who have mental illnesses from living in their neighborhood." While it may be illegal for them to discriminate against you, good luck proving it. You may also be denied certain licenses as a result. You can always lie, but they may find out and you may be fined or expelled.

If you're an employed TI, you will probably be Mobbed. You should know that EAPs work for the company and operate on the premise that if there is a problem, it's with you. If you complain about Mobbing or a similar phenomenon, they may recommend that you seek help from a mental health professional. As concerned as they may seem, they may suggest this so that you can be discredited later if necessary. According to Bruce Ennis, an ACLU attorney who represents people denied of jobs due to mental illness stigma, it's better to be an ex-convict in the job market than someone with a record of mental illness. It's like a prison sentence that follows you wherever you go.

Results of the Hidden Evil

Although at this point we have no way of knowing how many people this affects, McKinney estimates that at least many thousands are probably targeted. "There are persons I have seen being targeted ... [who are] completely unaware of what's happening," she says. "So those who are complaining of this are ... as I've said, just the tip of the iceberg. I would say this is probably very, very widespread."

Eleanor White, who has been targeted for over twenty years, and has documented this program on her website, www.raven1.net, sees this number as conservative. Based on a survey she conducted in 2005, she estimates that the numbers are probably in the millions. Consider that at one point in East Germany, 6 million people came under scrutiny by the MfS, out of a population of about 17 million. That's over one-third of the population. And mainstream media has told us that the U.S. is poised to recruit double the informants of East Germany.

At this point it is difficult to estimate the amount of misdiagnoses, suicides, miscarriages, deaths by cancer, and murders that this program has caused. But in addition to the known cases previously mentioned, I've included some below that occurred under curios circumstances. Over the last ten years there has been an increase in the amount of people "going postal" out in public and in the workplace. It is likely that the Hidden Evil has played a role in this increase. People that have been Mobbed have been known to go postal and kill themselves.

It would be very difficult to legally prove in a court of law that Mobbing and The Hidden Evil contributed to the events below, especially since there are currently no laws preventing this type of harassment. But, we now know that some agencies have carried out mind-control programs and systematic harassment in the past.

The stories that the victims describe are similar and the tactics are consistent with ones employed in covert harassment campaigns, such as complaints of recurrent harassment by multiple people, sabotage, setups, etc. The symptoms such as depression, accidents (other illnesses), and extended sick leave are also present. The symptoms victims experience result in nervous breakdown, violence directed at third parties, suicide, murder, and murder/suicide.

Beginning in 80s there was a wave of postal killings, giving rise to the term, "going postal." There have been about 24 separate incidents of postal violence since then including murder, suicide, and murder/suicide. Carefully read the evidence in this section. If you examine these events you may see a pattern in these seemingly senseless acts of violence. An identifiable pattern in these cases includes the use of EAPs, nervous breakdowns, followed by sick leave for psychological problems, and eventually suicide and murder.

It may not be a coincidence that these postal offices are fraught with horror stories of people murdering their family and coworkers. According to author Alex Constantine, former members of the intelligence community are often granted positions as postmasters. These offices are said to have been used as labs for behavioral modification, and psychological harassment on unwitting people. They have their own intelligence network within these offices, which target and harass people. This harassment sometimes continues no matter where they move to, or what environment they work in.

An article entitled *$5.5 Million Awarded in Sexual Harassment Case*, that appeared in the November 6, 1996 issue of *Employment Practices Solutions*, stated "A jury awarded $5.5 million to the family of a woman who was driven to suicide ... by her bosses and underlings at the Postal Service. They accused her co-workers of intentionally botching jobs or missing deadlines to sabotage her career." Apparently in an attempt to explain the covert violence she was experiencing, one of her actions included, "leaving a suicide note blaming the Postal Service."

Frequent harassment committed by multiple people, including superiors, coworkers, and subordinates, directed at one person, which results in suicide is Mobbing. Because the existence of The Hidden Evil and Mobbing are not yet widely admitted, this article was mislabeled. There is no existing category that targets of this violence fall into, other than being decent people. Fortunately her family was able to focus on an existing law and win the case. Did she experience public harassment too?

The Guardian ran an article on January 9, 2001 entitled, *Suicide of Black Worker Caused by Bullying.* It stated that "Mr. Lee left a suicide note addressed to his mother, Unnell, which said: 'Tell them it was nice playing with me and they won.'" This man's father wrote an open letter to 1,500 postal workers in Birmingham seeking more evidence about the harassment. "But instead of help, he received threats," stated the *Guardian.*

It continued, "Two other workers who claim they were bullied by colleagues are on long term sick leave," and one woman suffered a "nervous breakdown." Suicides, nervous breakdowns, and extended sick leave due to repeated harassment by multiple colleagues, all signify Mobbing. This also illustrates the *group mind* of staying loyal, and covering the system that keeps them employed, no matter how sinister.

Mobbing targets sometimes go on sick leave and are unable to ever return to the workforce. An environment of fear and danger is deliberately created, where nobody knows who the next victim will be. To get an idea of what this environment is like, picture a sinking ship where nobody knows if there will be enough room for them on the lifeboat. The people who oversee these operations are well aware that when people are put in an environment of lingering death, the *herd mentality* will be activated, and cause a struggle for survival. This brings out the worst in most people. This environment is deliberately created for their sick amusement and control.

According to an article labeled, *Report Links Violence, Local Postal Management*, which appeared in the December 22, 1998 edition of the *Milwaukee Journal Sentinel*, the post office may have been responsible for some of these murders. The article then went on to describe how programs specialist at the post office were attributing the killings to "random acts of violence," and that they were conducting their own version of an 800-page report which would not link the post office to the killings.

In 1999 a postal inspector named Calvin Comfort was assigned to investigate some of these postal killings. While doing so, he suffered a nervous breakdown and had to take a year off. After returning to work he killed himself in the parking lot outside the building. It's clear to me that this was a message. He essentially told us that they did it.

Covering this story, the *Milwaukee Journal Sentinel* ran an article on April 2, 1999, entitled, *Postal Inspector Kills Self Near Work.* The article stated, "A Milwaukee postal inspector who investigated the fatal 1997 shooting spree at the downtown post office took his own life outside the same building... Comfort's investigation into the shootings seems to have been at least partly responsible for a nervous breakdown he suffered..."

They continued, "Comfort was seen by a therapist available to employees through their employee assistance program [EAP]. A separate report, to which Comfort contributed, deals specifically with the events leading to the shooting and is still unreleased. Inspector Rudy Green, is quoted in the medical examiner's report as saying Comfort "has been under a lot of stress since ... he was the lead investigator in a homicide that occurred at the Milwaukee main post office." Under a lot of stress? I'm convinced he was violently attacked and murdered to prevent him from uncovering some incriminating evidence. Was he also targeted by the Hidden Evil? In my opinion, yes.

USA Today published an article on January 31, 2006, entitled, *Ex-Postal Worker Commits Suicide After 5 Die in Shooting.* It stated, "A former postal worker who had been put on medical leave for psychological problems shot five people to death at a huge mail-processing center and then killed herself in what was believed to be the nation's deadliest workplace shooting ever carried out by a woman."

The article continued, "The attack Monday night was also the biggest bloodbath at a U.S. postal installation since a massacre 20 years ago helped give rise to the term 'going postal.'" Adding evidence to the charge that these people already know who they're going to kill in advance, it ended with, "chances are she might have known her victims." Of course she did, they were harassing her.

Why would she go back to the post office to carry out these attacks after being on sick leave for two years? It's apparent to me that she took revenge where she believed her torment originated. My contention is that her psychological problems were the result of mental injury, and that when she left, she was never allowed to heal due to public harassment. They literally kept beating her until she broke.

But the real sick part is that those who have these individuals under surveillance during these harassment campaigns know exactly what they're planning—and *allow* it to happen. They cause it. "Psychological harassment and intimidation are carried out by hired specialists," wrote author Constantine. The "U.S. Postal Service and other federal agencies draw on their services."

The book *Mobbing* states, "Some may feel revengeful and direct their rage at the Mobbers. Seeking revenge for their misery, they may 'go postal.'" Regarding the lack of media coverage they conclude, "There is rarely a report that identifies the deeper background of these tragedies." McKinney arrived at a similar conclusion, and described a refusal of the mass media to address this topic.

These are just examples, and this phenomenon occurs in many institutions. If there was an actual unbiased investigation of suicides and murders caused by this program, the post office may not necessarily lead other institutions. The U.S. postal service has conducted its own $4 million dollar study to disprove the myth that postal workers are more prone to violence than other workers. One of their goals was to hopefully remove the phrase, "going postal" from the language of mayhem. Apparently their study concluded favorably that postal workers are no more likely to "go postal" than other types of workers.

Summary

The objectives of the program are apparently to destroy a person from the inside out, cause them a nervous breakdown, facilitate their hospitalization or imprisonment, and force them to commit murder or suicide. These objectives may be accomplished by the removal of economic and social support structures, combined with repeated public and electronic attacks. Those on the receiving end of public harassment may experience trauma as a result of repeated injury.

Electronic harassment may additionally cause a variety of problems, up to and including heart attacks, strokes, cancer, and aneurysms. Targets may be misdiagnosed, labeled mentally ill, or be persuaded to take medication for an illness that does not originate from within.

Volume II Part II

Chapter 24

The Psychopathic Influence

Description of a Psychopath

Both the financial elite and their servants who maintain this system, appear to exhibit behavior that is consistent with symptoms associated with a medical disorder known as *psychopathy*.(*) Psychopaths, also called sociopaths, are categorized as those who exhibit superficial charm and intelligence, and are absent of delusions or nervousness. Their traits include:

- Unreliability
- Frequent lying
- Deceitful and manipulative behavior (either goal-oriented or for the delight of the act itself)
- Lack of remorse or shame
- Antisocial behavior
- Poor judgment and failure to learn by experience
- Incapacity for love
- A poverty of general emotions
- Loss of insight
- Unresponsiveness in personal relations
- A frequent need for excitement
- An inflated self-worth
- An ability to rationalize their behavior
- A need for complete power
- A need to dominate others

Psychopathy is basically an emotional disorder. The book, *The Psychopath*, by James Blair, Karina Blair, and Derek Mitchell, states, "The crucial aspect of psychopathy is ... the emotional impairment." According to Dr. J. Reid Meloy's book, *The Psychopathic Mind*, although psychopaths don't feel emotion in a normal sense, they do experience boredom, envy, exhilaration, contempt, sadistic pleasure, anger, and hints of depression.

Generally, those who believe it's caused by environmental factors use the term sociopath, and believers of the biological theory use the term psychopath. Psychopathy closely resembles Antisocial

Personality Disorder (ASPD or APD) or Conduct Disorder (CD) as outlined in the DSM-IV. These disorders are detected using the Psychopathy Checklist-Revisited (PCL-R), the DSM-IV, and other diagnostics.

These character types, comprise about 4% of the population and span every level of society. Psychopaths can be found in every race, culture, profession and class. Because the term psychopath has been used to describe APD types and sociopaths, in this chapter I'll use it as a universal label for these three character types.

Later when I'm explaining how psychopaths always mask themselves when seeking positions of power, it will help to remember the following: If a rational person tries to apply their logic while trying to understand the reason for an objective or act of a psychopath, they will fail. This will be explained in more detail later. Likewise, when a rational person hears of the possibility that a massive lie has been told to a population by a trusted leader, and they attempt to use their logic to determine weather or not such a lie is possible, they will usually not believe the truth (that they have fallen for a huge lie).

The reason for this is that although most of us can identify with small lies, we find it difficult to conclude that such a massive lie is possible. When I use the term *massive lie*, I don't just mean a complete falsehood regarding a major event, but also the scope of its influence (global) and the amount of people that have fallen for it.

In his book, *The Mask of Sanity,* Dr. Hervey Cleckley, says that even during the most "solemn perjuries" they show "no difficulty at all in looking anyone tranquilly in the eyes." He adds that that they will "lie about any matter, under any circumstances." He explains that it is difficult to express how completely straightforward they appear when telling a blatant lie.

"The great masses of people will more easily fall victims to a big lie than to a small one."
-Adolph Hitler

"Lying, deceiving, and manipulation are natural talents for psychopaths," agreed Dr. Robert Hare, in his book, *Without Conscience*. "When caught in a lie or challenged with the truth, they are seldom perplexed or embarrassed—they simply change their stories or attempt to rework the facts so that they appear to be consistent with the lie."

Psychopaths are *always* able to justify their actions, no matter how brutal. They have, "an ability to rationalize their behavior so that it appears warranted, reasonable, and justified," says Dr. Cleckley. Dr. Hare added, "Psychopaths show a stunning lack of concern for the devastating effects their actions have on others. Often they are completely forthright about the matter, calmly stating that they have no sense of guilt, [and] are not sorry for the pain and destruction they have caused," which, says Dr. Hare, "is associated with a remarkable ability to rationalize their behavior."

Psychopathy is usually untreatable. Most therapists won't work with them because they often end up damaged in the process. Dr. Hare explained, "Such counseling would be wasted on psychopaths." Some of them will even reflect the wishes of the therapist and pretend to be getting better.

In his book, *People of The Lie*, psychiatrist Dr. Scott Peck had this to say: "Among themselves therapists will not infrequently refer to a patient's psychopathology as being 'overwhelming.' We mean this literally. We literally feel overwhelmed by the labyrinthine mass of lies and twisted motives ... into which we will be drawn if we attempt to work with such people..."

Wikipedia describes that, "traditional therapeutic approaches actually make them, if not worse, then far more adept at manipulating others and concealing their behavior. They are generally considered to be not only incurable but also untreatable." Basically psychopaths are the way they are for life. In most legal jurisdictions they are considered sane. So technically, they're not mentally ill, just different.

Dr. Scott Peck concludes, "I have learned nothing in twenty years that would suggest that evil people can be rapidly influenced by any means other than raw power. They do not respond," he says, "to either gentle kindness or any form of spiritual persuasion with which I am familiar with."

Where Are They?

When people hear the word *psychopath*, most think of the famous serial killers locked away in prison. However, most don't end up in prison or mental hospitals. Dr. Cleckley wrote, "The true difference between them and the psychopaths who continually go to jails or to psychiatric hospitals is that they keep up a far better and more consistent outward appearance of being normal."

"This outward appearance," says Dr. Cleckley, is essentially a mask, which, "may include business or professional careers that continue in a sense successful, and which are truly successful when measured by financial reward or by the casual observer's opinion of real accomplishment."

"Many psychopaths never go to prison or any other facility," agreed Dr. Hare. "They appear to function reasonably well—as lawyers, doctors, psychiatrists, academics, mercenaries, police officers, cult leaders, military personnel, business people, writers, artists, entertainers, and so fourth—without breaking the law." He continued, "Their intelligence, family background, social skills, and circumstances permit them to construct a facade of normalcy."

"Corrupt and callous politicians, social or career fast climbers, authoritarian leaders, abusing and aggressive persons, etc., are among them" wrote Dr. Renato Sabbatini in his article, *The Psychopath's Brain*. "A common characteristic," says Dr. Sabbatini, "is that they engage systematically in deception and manipulation of others for personal gain. In fact, many successful and adapted non-violent sociopaths can be found in our society."

Most of these people are not just right in your churches, schools, charitable organizations, and workplaces, but by their very nature, they are likely *running* them. It is a core trait of the psychopath to place themselves in positions of influence, not for public service, but for *power*. "The experience of pleasure is not reciprocal for the psychopath," stated Dr. Meloy, "it is available only through sadistic channels of power and control." Achieving power for the sake of having power is the nature of the psychopath. "They love to have power and control over others," agreed Dr. Hare.

The need for absolute power over others and the wish to inflict pain for the enjoyment of watching others suffer, are almost never apparent to the casual observer. The reason for this is that another core trait of the psychopath is *disguise*. So unfortunately, these individuals usually mask themselves as good-natured people. If they have tremendous wealth, you can bet that they'll create charitable organizations as part of their mask.

They are well aware that their mental makeup is drastically different from the majority. They have a sixth sense for detecting and exploiting *any* weakness you may have. At a very early age they learn that they can inflict mental and emotional harm on others with ease. They also learn how to detect others like themselves out of a crowd of normal people. Beginning in their childhood, most of them learn to mimic normal emotional reactions in order to blend in with society.

An article on Dr. Hare's website called, *Psychopaths Among Us*, by Robert Hercz, describes how Dr. Hare was contacted by Nicole Kidman, who wanted his advice on how to play the part of a psychopath for her film, *Malice*. Dr. Hare uses the anecdote of a psychopath who had just witnessed an accident where a mother watched her child get killed by a car. There's blood all over the place, and the psychopath experiences no emotion, but instead, is trying to avoid getting blood on her shoes. The psychopath notices the mother's emotional reaction to the accident and is fascinated. She goes home, looks in the mirror, and begins to mimic the facial expressions of the mother. "That's the psychopath," revealed Dr. Hare.

Wikipedia, the online encyclopedia, states that, "any emotions which the primary psychopath exhibits are the fruits of watching and mimicking other people's emotions." They are adept at, "using their charm and chameleonlike abilities to cut a wide swath through society and leaving a wake of ruined lives behind them," Dr. Hare warns.

"More often than not," says Dr. Cleckley, "the typical psychopath will seem particularly agreeable and make a distinctly positive impression when he is first encountered. Alert and friendly in his attitude, he is easy to talk with and seems to have a good many genuine interests. There is nothing at all odd or queer about him, and in every respect he tends to embody the concept of a well-adjusted, happy person."

"Psychopaths are often witty and articulate," concurred Dr. Hare. "They can be amusing and entertaining conversationalists, ready with quick and clever comeback, and can tell unlikely but convincing stories that cast themselves in a good light. They can be very effective in presenting themselves well and are often very likable and charming."

Remember, most of them don't psychically hurt people, so this is about mental and emotional domination. To accomplish these objectives, they will use their mask of sanity to place themselves in positions within your community. These positions may include school boards, charitable organizations, churches, politics, law enforcement, or any position which they believe will offer them power over others. These are the places where most psychopaths end up, not jail.

A Different Species

Some researchers agree that the traits exhibited by these people produce a division stronger than age, race, and religion, which places them in a new category of people. In other words, these people are almost not human as we know it. The word *antisocial* does not describe someone who prefers to sit at home rather than attend gatherings. More accurately it means *antihuman.* Most people can't bring themselves to understand the mind-set of a psychopath. Dr. Hare explained, "Imagining the world as the psychopath experiences it is close to impossible."

In his book, *Political Ponerology: A Science on the Nature of Evil Adjusted for Political Purposes*, Dr. Andrew M. Lobaczewski wrote, "The pathocratic world, the world of pathological egotism and terror, is so difficult to understand for people raised outside the scope of this phenomenon that they often manifest childlike naiveté, even if they have studied psychopathology and are psychologists by profession."

But, if you can, try to imagine someone who seeks power for no reason other than to have power, or someone who deceives just to experience the delight of having done so. Or someone who tortures another person physically or emotionally for the enjoyment of watching them suffer. Imagine someone doing these things, and not losing a moment of sleep at night (zero remorse). And add to all this, the ability to conceal themselves from an extremely naive public. Psychopaths are essentially aliens that look human. Probably one of the best skills to learn is how to detect and avoid a psychopath.

"The typical person, of whatever nationality, wants only to enjoy success in his job, to be able to afford a reasonably high standard of living complete with leisure and travel," describes Gary Allen. "He wants to provide for his family in sickness and in health and to give his children a sound education. His ambition stops there," says Allen. "He has no desire to exercise power over others, to conquer other lands or peoples, to be king."

He continues, "Since he has no lust for power, it is difficult for him to imagine that there are others who have [and] who march to a far different drum." He asks, "Why should we assume there are no such men today with perverted lusts for power?" In my opinion, we shouldn't because we know that they do exist, and consist of about 4% of the population during any period. "And if these men happen to be billionaires," ponders Allen, "is it not possible that they would use men like Hitler and Lenin as pawns to size power for themselves?" As I've demonstrated, true history supports this claim exactly.

Dr. Hare refers to them as "intraspecies predators." "There is a class of individuals who have been around forever and who are found in every race, culture, society and walk of life," he says. "If you think about it," he adds, "you will realize that what is missing in this picture are the very qualities that allow human beings to live in social harmony." "It is not a pretty picture," he warns, "and some express doubt that such people exist."

A 2005, report entitled, *Antisocial Personality, Sociopathy, and Psychopathy*, by North Carolina Wesleyan College, describes them as, "morally depraved individuals who represent the 'monsters' in our society. They are unstoppable and untreatable predators whose violence is planned, purposeful and emotionless."

In addition to carrying out goal-oriented acts, the psychopath will also deceive and create chaos for *no reason* other than the enjoyment of doing so. "He will," says Dr. Cleckley, "in fact, commit such deeds in the absence of any apparent goal at all." This critical factor is often the one which baffles most rational people. When seeking an explanation for the behavior of a psychopath, they will attempt to apply reason. But, when dealing with a psychopath, we must understand that psychopathy *is* the reason.

One day a scorpion and a frog were by a river's edge and both needed to get across to the other side. The scorpion said to the frog, "Mr. Frog, would you be so kind as to let me on your back as you swim across this river? I have important business to conduct on the other side and I cannot swim in such a strong current." The frog was a little perturbed and so began to question the scorpion's motives. The frog spoke, "Mr. Scorpion, while I can appreciate the fact that you have business to conduct on the other side of this river, please consider what you are saying."

"You are a scorpion. You have a large stinger at the end of your tail. As soon as I let you on my back you will proceed to sting me which I might add IS your nature." The scorpion, ready for this, replied, "My dear Mr. Frog, it is clearly not in my interest to sting you at all! I do need to get to the other side and I PROMISE that no harm will come to you." Well, this made a great deal of sense to the frog, so the scorpion crawled on his back to make the trek across the river.

The frog was making good time getting across when all of a sudden, in the middle of the river no less, the scorpion began to sting the frog repeatedly. The frog, shocked at this development, cried, "Why, oh why are you doing this? You said you needed to get to the other side to conduct your business!" The scorpion replied casually, "Mr. Frog, you said it yourself. I am a scorpion. I have a large stinger at the end of my tail. And yes, it is in my nature to sting you."

Psychopaths in Politics

If this 4% of the population exists at all levels of society at any given period, is it possible that some would seek top positions of influence? Of course. Remember, a need for *absolute power* over others is a core trait of the psychopath. Goal-oriented deceitfulness, superficial charm, an outward friendly appearance, and having no remorse, are other traits which will allow them to achieve their goals. If they are also people of tremendous wealth, they will definitely use this to further their objectives. And because deceitfulness is a core psychopathic trait, this will probably be done by creating a humanitarian front organization.

Evil people are often busy building for themselves various fronts for disguise and to further their ambitions. Dr. Peck described, "They are likely to exert themselves more than most in their continuing effort to obtain and maintain an image of high respectability." So, unlike in the movies, evil does not reveal itself as the bad guy dressed in black, or the monster in plain site. Evil will very rarely expose itself to public light. *It must hide*. And it almost always hides under the guise of something righteous.

In fact, rather than hiding in the shadows dressed in black, it disguises itself in suits, uniforms, and charitable organizations, which allow it to deceive us into thinking it's our savior. This enables it to cause far greater damage. "While evil may manifest itself obviously ... it rarely does so," Dr. Peck proclaimed, and added, "those who are evil are masters of disguise."

Now let's examine the track record of the people who control the planet. They have created wars for profit and control, which have resulted in the deaths of millions of people, for which they exhibit no remorse. They've allowed attacks to occur, created depressions, and overthrown governments to further their ambitions. They have repeatedly lied about these events using the media, which they control, and academia, which they also control by their Tax-exempt Foundations.

These wealthy people *masquerade* as the saviors of society. They *disguise* themselves using public front organizations, which *appear* to be humanitarian in nature. They've repeatedly not only lied about major historical events, but have deliberately created these catastrophes for their own benefit. Their own publications indicate that they plan to install a global dictatorship with them in complete control. These people are textbook psychopaths.(**) They are *antihuman*!

Compared to regular law-breaking criminals, these people appear to be OK, on the surface. But as we've learned, lies, deception, and disguise are standard traits of those who are evil. "[They] come from the very highest social strata," wrote Allen. "They are immensely wealthy, highly educated and extremely cultured. Many of them have lifelong reputations for philanthropy. Nobody enjoys being put in the position of accusing prominent people of conspiring to enslave their fellow Americans, but the facts are inescapable."

If you were to research the people throughout history that have committed atrocities against populations, you would probably find that most were psychopaths. Although mainstream history books do describe some of these events, they don't emphasize the pattern which they were part of. This pattern consists of a steady stream of psychopaths lying, deceiving and murdering their way to the top, which then resulted in the atrocities. This re-occurring theme seems to be the norm.

Regarding this, Dr. Lobaczewski said, "We then usually reach the conclusion that the deed would not have taken place ... since the pathological factor sealed its occurrence or became an indispensable component it its origin. The hypothesis thus suggests itself that such factors are commonly active in the genesis of evil." He explained, "Within this interlocking structure, one kind of evil feeds and opens doors for others regardless of any individual or doctrinal motivations." Finally, he suggests, "Since pathological factors are present within the syntheses of most instances of evil, they are also present in its continuum."

"Because their willfulness is so extraordinary—and always accompanied by a lust for power—I suspect that the evil are more likely than most to politically aggrandize themselves," described Dr. Peck. Dr. Sabbatini wrote, "Under stressing social situations such as in wars, general poverty and breakdown of the economy, sweeping epidemics or political fighting, etc., sociopaths may acquire the status of regional or national leaders and saviors... When they are in positions of power, they can inflict far more damage than as individuals."

Professor Marrs commented, "Yet, in public, they impress us as men who are likeable, intelligent, refined, fair-minded, tolerant, thoughtful, kind, and gentile men who sincerely care about such matters as the environment, the plight of the hungry and starving overseas, the jobless, and the poverty-stricken. Moreover, they often are recognized as leaders in the legitimate quest for world peace and tranquility." He continued, "Many are active in church work and charitable organizations. Others give freely to good causes." Recognizing this facade, he added, "No one would suspect for a moment what actually goes on in the deep, dark recesses of their diabolical minds."

Another extremely useful skill to learn is the ability to detect this pattern of decay in a nation under a pathological influence. Dr. Lobaczewski's book, *Political Ponerology* explains this pattern in detail. He has experienced this process personally while living in Poland under both Nazi and Communist occupation.(***) Once you realize that psychopaths control America and the NATO nations, the reason for endless wars and never-ending turmoil becomes clear. Once you understand the nature of the psychopath, it makes perfect sense.

When psychopaths rule a society, it will exhibit *their* traits. Generally it will be heavily corrupted. But because deception is a primary trait of the psychopath, it will appear humane. Its traits can be observed from the highest level of government down to the street level. If you wanted to witness the madness of a civilization under psychopathic rule, you need not visit the state capital or a major city, because even the smallest village will exhibit these traits.

When the controlling faction of a society is evil the sickness permeates into the lower levels. Immoral attitudes are projected, while moral ones are ridiculed, and enemies are created. The destructive process ends with a holocaust, genocide or persecution of some manner. The *New World Order* is such a movement.

In November of 1975, Congressman Lawrence P. McDonald stated that the most important issue of our time is the "drive of the Rockefellers and their allies to create a one-world government, combining super-capitalism and Communism under the same tent, all under their control." He warned that this drive is "international in scope," and "incredibly evil in intent."

Ross wrote, "I don't know about you, but these ... activities seriously concern me, because my children and grandchildren will suffer many times greater than we do today under the control of these EVIL MONSTERS." [Emphasis in original] He added, "I have tried to find worse terms for them, but this is the best that I can think of to describe them."

When a nation or other organization begins this process of evil, psychopaths and other deviants are attracted to it like a magnet. Like-minded individuals are installed in key positions of influence. Regarding the current epidemic, this would include the federal agencies which are allegedly carrying out these Cointelpro operations and hitting people with Non-lethal Weapons, as well as the citizen networks which stalk targets in public.

As already demonstrated, you must capture the streets and a percentage of the population in order to install a dictatorship. To do this, a portion of the society is selected to enforce the dictator's rule on the street level. Historical evidence suggests that this portion usually consists of the core operational centers of a society, e.g., factories, hospitals, schools, civic centers, religious organizations, law enforcement, utility companies, stores, etc.

Dr. Lobaczewski described this in the following way: "The actions of this phenomenon affect an entire society, starting with the leaders and infiltrating every village, small town, factory, business, or farm. The pathological social structure gradually covers the entire country, creating a 'new class' within the nation. This privileged class of deviants feels permanently threatened by the 'others.'" The "others" are apparently the people who have been targeted for persecution.

The Hidden Evil

It's very difficult to describe with words an event that must be experienced in order to understand. The event I'm referring to is the observation of the psychopath or psychopathic system in its true form, after the mask has been removed. It can be described as something *deeply horrific*. This is a gut-wrenching experience for a normal person, and can cause severe trauma, especially if they can't escape it.

"If someone has personally experienced such a nightmarish reality," says Dr. Lobaczewski, "he considers people who have not progressed in understanding it within the same time frame to be simply presumptuous, sometimes even malicious." "This experience," he continues, "[is] unceremoniously rejected by ... [people and] becomes a psychological burden for him, forcing him to live within a narrow circle of persons whose experiences have been similar."

This encounter with a psychopathic element has been the experience of many Mobbing and Hidden Evil targets across the planet. McKinney wrote that one objective of these harassment campaigns is to instill a perverted sense of loyalty toward the agency responsible for the harassment. This makes perfect sense when the pathological factor is considered. "Sadistic control is also an element of perversion," wrote Dr. Meloy, which is an indication of "psychopathic disturbance."

McKinney also mentioned that these agencies appear to treat their targets as objects while harassing them. "Psychopaths view people as little more than objects to be used for their own gratification," explains Dr. Hare. "The weak and the vulnerable—whom they mock, rather than pity—are favorite targets." This helps to explain the reports of people being targeted by the Hidden Evil for no particular reason, other than being decent. A psychopathic program will target such people.

Just as the creators of this program are masked as humanitarians, the program itself is masked as beneficial and necessary. Naive individuals who participate, may not immediately recognize this. But a percentage of those who control the weapons which torture civilians (it takes a psychopath to do this), as well as those on the lower levels who follow and harass targets, are psychopaths. And they're aware of the true nature of this program. That's why the joined.

"The outer layer closest to the original content is used for the group's propaganda purposes," says Dr. Lobaczewski. "Average people succumb to the first layer's suggestive insinuations for a long time before they learn to understand the second one as well." "Anyone with certain psychological deviations," He explains "especially if he is wearing the mask of normality ... immediately perceives the second layer to be attractive and significant; after all, it was built by people like him."

The pathological factor also helps to clarify why people who are, arguably, more spiritually and morally evolved than the masses, are being identified and targeted. According to Dr. Lobaczewski, these types of individuals are the first to be targeted in regimes controlled by psychopaths. Dr. Meloy explained that the psychopath's "perception of others' pleasures arouses only envy and greed in themselves." And that they receive, "gratification of sadistic impulses through intentional infliction of emotional or physical pain upon others."

"Psychopathic individuals who never enter psychotherapy are paradigms of this hatred of goodness," says Dr. Meloy. According to Dr. Meloy, although envy is not consciously felt, it is the driving force for the motivation of their destructive behavior. He describes this behavior as "manipulative cycling" which includes a "mocking, [and] controlling attitude" while attacking their targets.

"The manipulative cycle," explains Dr. Meloy, "both enhances his narcissism and protects his vulnerability." This cycle is ongoing because the threat to the inflated self is ever present. He describes this cycle as a purification process for the psychopath. "The desire to control and degrade the actual object . . . may be fueled by the sadistic pleasure inherent in the behavior," he says.

The mocking and controlling behavior exhibited by these federal agencies against their targets, appears to be part of the manipulative cycling which Dr. Meloy speaks of. This makes even more sense when we consider Dr. Lobaczewski's premise, which states that the controlling psychopathic faction of a society (the financial elite), will recruit lower-level psychopaths to do their bidding. These lower-level deviants naturally seek employment in law enforcement, security, the military, politics, or other positions which they believe will offer them power.

Summary

A small portion of the population have a psychological makeup which is much different than most. They are completely aware of their difference. They also know that most people are not aware of this profound separation. The difference includes an emotional deficiency, accompanied by a lack of remorse, which allows them to operate outside of standard moral boundaries. They are able to conceal this difference to some degree and usually appear to be generous and friendly.

They consistently engage in antisocial behavior which includes destroying people's lives, in order to feed their inflated egos. During this process they frequently enjoy mocking their targets, which they see as weak, or are envious of. They will inflict pain upon others for no reason other than the enjoyment of doing so. They span all levels of society. Some authors consider them to be a different species of humans.

Psychopaths naturally gravitate toward positions of power. Many historical atrocities were caused by psychopaths. An organization or nation under a psychopathic influence will become saturated with its sickness. It will exhibit these destructive traits, from the upper levels, down to the smallest village. The Hidden Evil is a psychopathic program which exists in a society controlled by psychopaths.

Endnotes

* It's nice to know that there's an actual scientific category that these people can be placed in. I first learned that psychopathy was likely the specific illness that the creators of this program suffer from while listening to author and radio host Alan Watt at www.CuttingThroughTheMatrix.com. He now hosts a show on the *Republic Broadcasting Network*.

** This charge is based on my unprofessional opinion. A true diagnosis is a complex matter involving an in-depth diagnostics using the PCL-R, or DSM, as well as input from friends, family and employers.

*** Ponerology is derived from the Greek word *poneros*, which means evil. Political Ponerology is the study of the growth of political evil. According to the author of *Political Ponerology*, Zbigniew Brzezinski of the Trilateral Commission attempted to prevent the book from being published.

Chapter 25

The Satanic Influence

There is a definite Satanic manifestation within this program which is evident during group stalking, and more generally, by the true sinister nature of the program, which will be abundantly clear to most after careful study. The ongoing life-destroying tactics, accompanied by repeated messages intended to mock targets, are surely acts perpetrated by those who are influenced by evil. McKinney found that there is a striking similarity between the tactics used by this worldwide group and those reported by people who have been stalked after breaking ties with Satanic cults.

Dr. Kilde too, noted that “Satanism” is a common theme used by worldwide Gang Stalking crews. She wrote that group members “get rewarded for their evil actions,” which includes Satanic symbolism such as wearing yellow, orange, and black colors.

Targets may be stalked by an unusual amount of red vehicles and citizens dressed in red. These vehicles, which may be red or black, may cut targeted people off on the road, block them in a parking lot, or perform some other tactic to reveal their presence. License plates which include the numbers 666, 696, 616, and variations of these are used. Apparently these plates can be made-to-order directly through the DMV. McKinney wrote of the “use of modified license plates,” which “appears to have been acquired through the State DMV channels, thus suggesting government/intelligence agency involvement.” Citizens with Satanic apparel such as devil figures on shirts or other clothes may stalk targets.

The people who are used to convey these messages may not be aware that they're participating in a Satanic theme. This is not surprising because most appear to be unaware of the nature of the program in general. There may be several motivating factors used to gain the complicity of well-meaning participants. Those with red clothing or vehicles who surround targets may be told that this is necessary so the target will know that they are part of the team and will be persuaded to leave the community.

For symbolism that is more overt, such as devils and the number 666, participants may be told that if a target believes members of the community are actual Satanists, they’ll leave. Although I think the majority are not, some of these people may be Satanists. Multiple researchers contend that Satanists usually appear to be religious and are often much respected members of society. Consider the account of Dr. Peck in his book, *People of the Lie*; evil needs to hide, and almost always masks itself as righteousness.

The reason for the existence of this program as told to the citizen squads, stores and restaurants is a lie. The role they believe they play is a lie. Lies are spread about individuals to gain cooperation. So there are lies within lies. The program is literally, one... big... lie... As demonstrated, the consistent track record of those likely responsible for this secret program forms a clear picture of their true nature. I've already demonstrated the atrocities they have caused. But add to that, the lies, upon lies, upon lies, upon...

"Lying," wrote Dr. Peck, "is simultaneously one of the symptoms and one of the causes of evil, one of the blossoms and one of the roots." He says, "The evil in this world is committed by the spiritual fat cats ... by the Pharisees of our own day." "They are," he announced, "for instance, in my experience, remarkably greedy people. Thus they are cheap—so cheap that their 'gifts' may be murderous."

Satanism encompasses the international child sex slave trade, drug trafficking, human sacrifice, murder, and other crimes. In her book, *The New Satanists*, Linda Blood wrote, "Satanism is the fastest growing underground criminal movement in the world today" and that it is "directly linked to an exploding number of child abuse cases, Nazism, [and] drug dealing." Linda's analysis of Satanism harmonizes with the conclusions of other researchers, such as former Senator John DeCamp, and Ted Gunderson.

She adds, "Satanic crime—in fact, all violent, destructive, and criminal activity associated with malevolent forms of occultism—exists as part of a wider criminal continuum and must be understood and recognized within that context." Linda draws from her personal experience in a Satanic cult known as the *Temple of Set*, run by former army intelligence officer and psychological warfare expert, Dr. Michael Aquino.

Gunderson announced, "The video *Conspiracy of Silence* documents that children were flown to Washington D.C. ... for sex orgies with congressmen and senators." He contends, "My Finders report documents that the CIA was involved and ... still is involved in an international child-kidnapping ring. I have filed formal complaints with the FBI concerning this but they refuse to investigate it." Mainstream publications such as the *Los Angeles Times* and the *Philadelphia Enquirer* showed that the *Conspiracy of Silence* documentary was scheduled to air on the *Discovery Channel* on May 3,

1994, but before it did all copies were purchased and destroyed. Fortunately an anonymous person sent Gunderson a copy and it is available at his website, www.Tedgunderson.com.

But where is this Satanic influence coming from? Once again, if we follow the trail, we find it leads to the supra-governmental organizations previously mentioned, which are interlocked with Satanic cults. More evidence for this will be provided. "Satanists control and dominate the big corporations ... in America," proclaimed Gunderson, and added, they have "overtaken and now control for the most part all aspects of our commercial system..." He says there is a "satanic cult ... network operating within the United States, with ties to top level officials—including politicians, judges, law enforcement, U.S. intelligence agents, and leading professional business men and women." He continues to list specific organizations under a Satanic influence such as:

- The military
- Many law enforcement agencies, including the police
- Courts
- The Federal Reserve and Wall Street
- Education
- The AHA, APA, AMA, ADA
- Hospitals and mental institutions
- The FDA
- Communications such as internet, telephone and postal service
- Transportation such as airlines, trains, roads/highways
- Religion, science and research
- Utilities such as gas, electric, oil
- The entertainment industry including movie production, TV, music industry and casinos
- The mainstream media
- Most major corporations

This Satanic influence within the controlling elite organizations filters down into the executive branch and intelligence agencies. And from there it extends into state and local governments. Although there may be other organizations which are Satanic in nature that influence governments, I've focused on the *Bohemian Grove* and *The Order of Skull and Bones*. The existence of these organizations is incontrovertible. They are both heavily interlocked with the Think

Tanks. It is only a question of whether or not they are Satanic in nature and how much influence they have on our society.

The Order of Skull and Bones

The existence of The Order is irrefutable. Also called, *Skull and Bones, The Order of Death, The Brotherhood of Death*, and *The Order of the Deaths-head.* The Order was founded in 1832 by William Huntington Russell and Alphonso Taft. The first graduating class was in 1833. It was formed after Russell returned from a trip to Germany before his senior year. It is said to be an American branch (chapter 322) of a German Masonic/Illuminati group in Bavaria.

Confirmation of its existence and the fact that its members occupy key positions in business and politics have appeared in mainstream media. For instance, during the 2004 presidential candidacy, articles appearing in *The Baltimore Sun*, *CBS News*, *The Washington Post*, *The Guardian*, and *The Atlanta Journal-Constitution* confirmed that both Kerry and Bush were members.

Kerry and Bush also admitted being members on television during a live interview on *Meet the Press*. There have also been books published on The Order that have received favorable Establishment reviews. Although most of this mainstream coverage was a whitewash, it did reveal a pattern of members occupying the executive branch.

Each year during commencement week 15 juniors receive a "tap" and are asked "Skull and Bones. Accept or reject?" Most say yes, says Professor Sutton, author of *Americas Secret Establishment*. After initiation the "patriarchs" spend only one year at Yale, and upon graduation, they reportedly receive $15,000 and a grandfather clock. "The organization," states Professor Sutton "is oriented to the post graduate outside world." Members of The Order are elevated to key positions in government.

It was legally incorporated as the Russell Trust at Yale University in 1856. It conducts yearly meetings at Deer Island (sic) in New York, which was donated in 1906 by a member of The Order, G. Douglas Miller (1870). The core membership consists of about 20-30 wealthy American families who originally settled on the East Coast in the 17th century. These families include, Whitney, Lord, Phelps, Wadsworth, Allen, Bundy, Adams and others. Its membership also includes families which later acquired wealth, such as Rockefeller, Sloan, Pillsbury, Davison, Harriman, and Payne.

An article appearing in *The Atlanta Journal-Constitution* on March 6th 2004, entitled *Yalies Bush, Kerry Can Keep a Secret*, stated, "Bush and Kerry are only the latest Bonesmen to star on the national stage. President George Bush, the incumbent's father, was also a member of Skull and Bones..." The article continued to name other prominent individuals. Notable members of The Order include:

- Henry Luce (1920), *Time Magazine*
- Henry Lewis Stimson (1888) Secretary of State, Secretary of War to Truman, recommends Atomic Bomb be dropped on Japan and worked with General Marshall and FDR to allow Pearl Harbor to occur.(*)
- William H. Taft (1878), Supreme Court Justice and President
- William F. Buckley Jr. (1950), *National Review*
- Pierre Jay (1892), first chairman of the Federal Reserve
- Prescott Sheldon Bush (1917) worked with W. Averill Harriman to fund Nazis and Communists via Union Banking Corporation and Guaranty Trust Company
- Winston Lord (1959) Lord, Day and Lord, Law Firm
- William Putnam Bundy (1939) CIA, editor of CFR's *Foreign Affairs* magazine, Assistant Secretary of State
- McGeorge Bundy (1940) National Security Advisor, President of the Ford Foundation
- W. Averill Harriman (1913) Director of Guaranty Trust, Special Assistant to President Truman, funded Nazis and Communists
- E. Roland Harriman (1917) Brown Brothers, Harriman, director of Union Banking Corporation

An examination of these names reveals cross memberships in the CFR, TC, BB, and the Bohemian Grove. For instance, William F. Buckley Jr. was a member of The Order (1950), the Bilderbergers, the Council on Foreign Relations, and the Bohemian Grove.(*1) "Columnist William F. Buckley is on the CFR's membership rolls," wrote Professor Marrs. Buckley ... is a Skull and Bones alumni, [and] has also been honored as a participant in the secretive Bilderbergers Group." In his book, *Bohemian Grove: Cult of Conspiracy*, Mike Hanson wrote of The Order, "You will also notice that many of the names ... show up in Bohemian Club member lists." These two organizations, in turn, are interlocked with the Think Tanks.

George H. W. Bush was a member of The Order (1948), the Council on Foreign Relations, the Trilateral Commission, and was appointed U.S. Ambassador to the UN. Perloff acknowledged, "George Bush was a Skull and Bonesman, a director of the Council on Foreign Relations, and a member of the Trilateral Commission." He was also the director of the CIA and eventually became President of the United States. In addition to being a member of The Order, Professor Marrs concludes that Prescott Sheldon Bush was a founding member of the Council on Foreign Relations. George W. Bush, was initiated into The Order (1968), and eventually became president. Both George H. W. Bush, and George W. Bush have attended the Bohemian Grove.

Winston Lord became the chairman of the Council on Foreign Relations in 1983 and is also a Bilderberger. Professor Marrs wrote that "Winston Lord" of the "CFR elite is a Skull and Bones man" and is on the "steering committee of the Bilderbergers." Both Bundy brothers were CFR members and William was once the director. Henry Lewis Stimson was also a CFR member. E. Roland Harriman was a CFR member. W. Averill Harriman, who financed Nazis and Communists, was a member of The Order, and the Council on Foreign Relations. He also became the U.S. Ambassador to the USSR and Great Britain, as well as Governor of New York.

In more recent years we find that during the 2004 presidential campaign John F. Kerry (1966), George W. Bush and Ralph Nader's lawyer, Donald Etra (1968), were all members of The Order. This was not a coincidence. "The Order has been called a "stepping stone" to the Council on Foreign Relations, Bilderbergers, and the Trilateral Commission," wrote author Marrs.

Another example of this is offered by Jim Tucker's analysis of the 1992 election. He wrote "Bush had been a longtime member of the Trilateral Commission which has interlocking leadership with Bilderberg. Clinton had been a Trilateralist for seven years and was promoted to the Bilderberg in 1991. Thus the world shadow government owned both presidential candidates in a typical win-win race."

Of the initiation ritual, Professor Sutton concludes, "Even with our limited knowledge of the internal ritual of The Order we can make ... definite statements about the links between The Order and satanic beliefs." Within the ceremony, he says, are "aspects notably satanic." He explains, "The first link is through photographic evidence of the association of Skull and Bones with satanic devices, i.e., the skull and crossed bones. The second link is through satanic symbolism."

"In brief," continues Professor Sutton, "the photographs reveal the men portrayed as grave robbers who reject human dignity and decency and use satanic devices." Professor Sutton also notes the activity of "Patriarchs dressed as skeletons" that "howl and screech at new initiates." A *History Channel* documentary entitled, *Secret Societies*, hosted by Arthur Kent, added that the shortest senior is appointed "little devil," and wears a Satanic costume.

Commenting on the bizarre rituals and beliefs of the Order, Professor Sutton declared, "The reader may consider this juvenile, and it may well be." However, he adds, "On the other hand, these 'juveniles' are the men today running the United States." To that, I would add, the planet. Professor Marrs concurred, writing, "I agree. It is absurd." But, "I can only tell you that this is exactly what goes on."

Another consideration is whether these initiation rituals are significant after graduation. Evidence suggests that the answer is yes. Like the "cremation of care," which will be covered shortly, members of The Order are pledging their allegiance, not to a country or Jesus Christ as they would have you believe. The rituals are enacted for the purpose of worshiping a Satanic entity, and diminishing the conscience of the individuals who partake in them.

Similar to the Cremation of Care at Bohemian Grove, this diminishing process is necessary so they can continue to perpetrate heinous acts on citizens of the world. The "pledge" is essential so that members will never stray from this one-world movement and the Satanic aspect of it. "It is a psychological conditioning process," observed Professor Marrs, and "also an intense form of peer pressure ... a type of group training."

As elite and influential as members of this society are, they too are manipulated and are under a form of mind-control. They are being directed. Professor Sutton observed, "What happens in the initiation process is essentially a variation of brain-washing ... through heavy peer pressure" they become "prepared for a life of the exercise of power and a continuation of this process into future generations." He adds, "The ritual is designed to mold establishment zombies, to ensure continuation of power in the hands of a small select group from one generation to another."

"Like their counterparts, the Skull and Bones Society, the Bilderbergers try to keep what goes on at their meetings hidden from public scrutiny," wrote Professor Marrs. "These two clandestine organizations ... have much to hide or else their procedures would not be kept behind closed doors. He warned, "What these cynical and powerful men are doing to damage America, [and] take away our freedoms ... is criminal."

The Order has, says Professor Marrs, "been able to gain impressive positions of influence in ... the Trilateral Commission ... and the Council on Foreign Relations." Later he adds, "They have also been active in places of leadership for the ... foundations—Ford, Rockefeller, Carnegie, Russell Sage, etc." He explained, "The Order's current lock on the White House, gives the Brotherhood all the authority it needs to carry out sweeping changes in the international system of governance."

Professor Sutton also contends that The Order has infiltrated the top of public organizations such as the church, law, communications, industry, finance, and politics. For instance, the "American Historical Association, the American Economic Association, the American Chemical Society, and the American Psychological Association were all started by members of The Order or persons close to The Order," stated Professor Sutton. His list also includes, Education, Law, Politics, Economy, Psychology, Philanthropy, Medicine, Religion, and Media. He reveals, "These are key associations for the conditioning of society." Many of these organizations that the professors speak of were also mentioned by Gunderson.

One way The Order controls these major institutions is through infiltration, another is by founding them. Professor Sutton explains a phenomenon where The Order sets up organizations, then fades away. "One observation is that The Order gets the ball rolling in new organizations, i.e., puts [in] the FIRST President or Chairman, he explained, "and then, when operations are rolling along, often just fades out of the picture." Professor Marrs wrote that The Order has "infiltrated every aspect of society. In four areas in particular..." He lists these areas as, "American foreign policy," "Wall Street," "education, and religion."

The Order being first on the scene is evident especially among foundations and think tanks according to Professor Sutton. Noting specific acts of infiltration, he writes, the first "President of the Carnegie Institution ... was Daniel Coit Gilman [1852], but other members of The Order have been on Carnegie boards since the turn of the century." He adds, "Gilman was on the scene for the founding of the Peabody, Slater, and Russell Sage Foundations." And that, "McGeorge Bundy was President of the Ford Foundation from 1966-1979." In 1969, James Jeremiah Wadsworth (1927) formed the Peace Research Institute, which later merged with the Institute for Policy Studies (IPS). Andrew Dickson White (1853) was the first president of

Cornell University, and Daniel Coit Gilman was the first president of John Hopkins University.

There can be no denying that members of The Order are appointed key positions in government and industry. Now an important question would be: What are their plans? The activities of The Order are directed towards bringing about a "New World Order," warned Professor Sutton, which "will be a planned order with heavily restricted individual freedom, [and] without Constitutional protection..." He adds, "We deduce this objective by examining and then summing up the actions of individual members: there has been a consistent pattern of [this] activity over one hundred years."

Bohemian Grove

The existence of Bohemian Grove is also incontrovertible. There have been several major sources of grove information by first-hand experience. Philip Weiss of *Spy Magazine* infiltrated the grove in 1989 for about 60 hours. Professor Peter Phillips of Sonoma State University published information on the grove after he was invited to the summer encampment of 1994. In 2000 independent filmmaker Alex Jones and TV Producer and author Mike Hanson infiltrated the grove with a hidden camera. They produced a film of their experience called, *Dark Secrets Inside Bohemian Grove*. In addition, Mike Hanson authored a book called, *Bohemian Grove: Cult of Conspiracy*.

Other sources of the club's history include the California Historical Society, University of California Bancroft Library, the California State Library, the Bohemian Grove Action Network (BGAN), and a few mainstream media sources such as *The Washington Post*, the *New York Post*, and the *Sacramento Bee*. Most of these mainstream publications are sanitized whitewashes that divert attention from the occult activities which unfold there. Although I'll be primarily referencing some of the previously mentioned sources, I'll also include the personal experiences of others who have been there, as well as some interpretations of grove activities by people I consider to be experts in history.

Bohemian Grove is the summer retreat of the Bohemian Club of San Francisco. Mike Hanson wrote, "The Bohemian Grove is more properly known as the Midsummer Encampment of the Bohemian Club of San Francisco. It takes place annually each Ides of July in a majestic redwood grove outside Monte Rio, California..." The Bohemian Club of San Francisco was formed in 1872-3, as "as a gathering place for newspaper reporters and men of the arts and literature," wrote Professor Phillips.(*2)

The club is invitation only. According to Hanson and Phillips, although the club was formed by artists, in the 1880s it was eventually dominated by business elites. Some early members included, Mark Twain, Ambrose Bierce, Bret Harte, Wytter Bynner, Henry M. Stanley, Jack London, Bing Crosby, Charlie Chaplin, Will Rogers, and Douglas Fairbanks. Apparently the artists (men of talent) needed the financing of the industrialists (men of use), and these businessmen eventually took over the club. It is now a, "good-old-boys political network," described Hanson.

The Bohemian Club began camping out at various locations in Sonoma County beginning in 1878. Between 1893 and 1899 they rented Bohemian Grove, and in 1901 purchased about 160 acres. It now consists of about 2,712 acres of redwood forest. It is located in Sonoma County outside the small town of Monte Rio, situated along the Russian River, and is about 70 miles north of San Francisco. At the grove, there is a civic center with a Bar and Grill, Post Office, and store.

"While the Grove's public areas afford a communal atmosphere for the Grovers to socialize, the real action goes on in the private confines of the individual camps," stated Hanson. The Bohemian Grove is equipped with a fire department and medical care on the premises during the encampment. Wait staff and other workers must wear ID tags and are restricted to the dining area and the parking lot.

In the November 1989 issue of *Spy Magazine*, Philip Weiss wrote in an article entitled, *Inside Bohemian Grove*, that "No Trespassing" and "Members and Guests Only" signs cover the grounds, and guards with binoculars and infrared sensors watch the paths." Reportedly there are barbed wire fences surrounding it and helicopters patrol the perimeter during the encampment. Hanson commented, "The Grove is vigorously guarded during the entire encampment." The Sonoma County Sheriff's department is said to be the Bohemian Grove's own private security force. Professor Marrs' research also leads him to conclude that one of several layer of protection consists of sheriff's deputies. Compare this to the Bilderbergers, who are escorted by police and surrounded by SWAT teams during their meetings.

The encampment lasts for about two weeks, and it never rains during their stay. Lots of alcohol is available 24/7, and the food is apparently first-class. The average club member's age is reportedly 55. Between two and three thousand people attend each year.

Entertainment includes plays known as *Jinks*. The second Saturday of the encampment is referred to as the "Low Jinks," and the following Friday is the "High Jinks," also known as the Grove Play. There are apparently also Spring Jinks in June.

Bohemian Grove is basically a playground and policy-making vacation for the CFR, interlocking Think Tanks, and Tax-exempt Foundations. "The CFR and Bohemian Club membership lists reflect a lot of the same names," stated Hanson. "By recent comparison, I was able to count about 40 Bohemians who are also CFR members." He added, "If you're a member of the Trilateral Commission or the Council on Foreign Relations, chances are you're a member of the Bohemian Club, too."

In attendance are representatives of companies such as: Atlantic Richfield, Avery Dennison Corporation, Colgate-Palmolive, Calfed, CBS Broadcasting, IBM, Universal/Vivendi Entertainment Group, Mattel Incorporated, Media General Incorporated, Northrop Corporation, Philip Morris, Pacific Enterprises, Rockwell International, Sears and Roebuck, SCE Corporation, Security Pacific, Times Mirror Company, Union Bank, Union Electric Company, Bechtel Construction, Wells Fargo Bank, Southern California Edison, General Electric, Bankers Trust, and Bank of America. Former directors of the FBI and CIA have also attended. As well as representatives from CNN, Los Angeles Times, Time Warner, and Times Mirror corporation.

Hanson wrote, "Here, heads of state and industry, such as Henry Kissinger, Caspar Weinberger, James Baker, Dick Cheney, Malcolm Forbes, Stephen Bechtel and a host of prominent CEOs, foundation chairs and university presidents mingle. ... About one-fifth of the members are either directors of one or more of the Fortune 1000 companies, corporate CEOs, top governmental officials (current and former) and members of important policy councils or major foundations." He proclaimed, "99 percent of them are white. And all of them are very, very rich."

"The remaining members are mostly regional business/legal elites with a small mix of academics, military officers, artists, or medical doctors," added Professor Peter Phillips, in his article, *San Francisco Bohemian Club: Power, Prestige and Globalism*, which appeared in the *Sonoma County Free Press* on June 8, 2001. "I found at least 150 Bohemians hold nearly 300 directorships in the top 1000 U.S. corporations," said Hanson. "Of these, I spotted several corporations with more than one Bohemian serving as directors. These include Bank of America, PG&E, (Pacific Gas and Electric), AT&T, Ford Motor Company and General Motors."

Notables that have attended the grove include, William F. Buckley Jr., George H. W. Bush, George W. Bush, Jimmy Carter, Bill Clinton, Walter Cronkite, Dwight D. Eisenhower, Queen Elizabeth, Leonard Firestone (Firestone Tires), Henry Ford II, Gerald Ford, Mikhail Gorbachev, Herbert Hoover, Robert Kennedy, Henry Kissinger, Richard M. Nixon, David Packard (Hewlett Packard), Prince Philip, Dan Quayle, Ronald Reagan, David Rockefeller, Laurence Rockefeller, Nelson Rockefeller, Theodore Roosevelt, Donald Rumsfeld, Helmut Schmidt (former German chancellor), William Howard Taft, Colin Powell, and Newt Ginrich.

The grove consists of a variety of camps which members are assigned to depending on their status. Reportedly, at one point the grove had over 200 camps, but now has about 116. The most prestigious camp is *Mandalay*. "*Mandalay*," wrote Hanson, "is the campsite where you'll find Henry Kissinger, former president Gerald Ford, Nicholas Brady, [and] George P. Shultz." Some camp names include: *Dog House, Isle of Aves, Pink Onion, Oz, Lands End, Rattlers, Red Fire, Jinks Band, Dragons*, and *Druids*. Former President Nixon belonged to the *Cave Man* camp, Reagan was a member of the *Owl's Nest*, David Rockefeller belongs to the *Stowaway* camp, and Art Linkletter is a member of the *Dragons* camp.

"The Bushes call the *Hill Billies* camp home," wrote Hanson, "as did Walter Cronkite, William F. Buckley Jr. and Christopher Buckley." He continues, "At *Owl's Nest*, Reagan camped out with the chairman of American Airlines, while over at *Hill Billies*, his vice-president George Bush Sr. was rubbing shoulders with the president of Eastern Airlines." "*Hill Billies*," he says, "also houses directors from General Motors, Southern Pacific, Westinghouse Electric, B.F. Goodrich, Morgan Guaranty Trust, Mutual Life Insurance, Metromedia, and Superior Oil."

Hanson asks, "Do the Bohemians really expect us to believe that ... cabinet officials, corporate chieftains and foundation directors ... share the same intimate camp site in the redwoods, and all these gentlemen are going to do is talk about music and the arts?" "There are few rules," wrote Weiss, "the most famous one being [the clubs motto] 'Weaving Spiders Come Not Here,'—in other words, don't do business in the Grove." But, he adds, the "rule is widely ignored." The club's motto was taken from Act 2 of Shakespeare's, *A Midsummer Night's Dream*.

"The media figures attending the retreat all agree not to report on what goes on inside," stated *Extra! Magazine*. One example of this media blackout occurred, "When Dirk Mathison, San Francisco bureau chief for *People Magazine* infiltrated the exclusive Bohemian Grove retreat ... [and] got a view into the U.S. elite that very few reporters have glimpsed." "Unfortunately," they revealed, "that elite includes the management of Time Warner, the owner of *People*, which prevented Mathison from telling his story."

They continued, "In 1982, *NPR* got a recording of Henry Kissinger's speech at the Grove—but declined to air it... Also in 1982, a *Time* reporter went undercover as a waiter in Bohemian Grove; like Mathison's *People* article, his story was killed." Hanson's experience is similar, he said, "After Alex and I returned from the Grove, Alex went on his radio show and told the world what we had done. So the next thing you know, we're getting calls from all these newspapers expressing interest," but "an editor above their heads would kill it." "Media access to the Grove continues to be limited," confirmed *The Daily Reveille*.

One topic of concern over the years has been the Lakeside Chats, also known as Lakeside Talks. Like topics covered by the Think Tanks, these topics become policy regularly. This is policy passed without the voting public's knowledge or approval. Lakeside Talks begin at 12:30 and last for about 30 minutes. "Arguably, the Lakeside Chats are the most important aspect of what goes on inside the Grove," wrote Hanson, and added, "matters of public interest are discussed and perhaps even decided here through the collective consensus-building process."

In their August 2, 1999 article entitled, *Movers, Shakers From Politics, Business go Bohemian*, The *Sacramento Bee Correspondent* stated, "The club has drawn criticism for years because of its emphasis on privacy." They added, "What particularly concerns Phillips and others are the "Lakeside Talks" held during the summer retreat."

One vocal group that has protested since 1980 against this type of activity is the Bohemian Grove Action Network (BGAN), run by Mary Moore. The *Sacramento Bee* announced, "The point of the protests, Moore said, has been 'to let the American public know that what they've learned in civics isn't the full story on how decision-making ... is made in this country.' The Bohemian Club, she said, 'is one of the most elite organizations on the planet.'"

The article continued, "Phillips echoes Moore's objections to the off-the-record nature of the Lakeside Talks. 'These are extremely powerful people and private discussions on policy issues that affect us certainly go against democratic principles. There's no reason that those speeches they're giving couldn't be transcribed and made public. They have a responsibility to be open about it.'" Weiss wrote, "The important men come out for the Lakeside Talks, and each speaker seems to assume that his audience can actually do something about the issues raised, which, of course, it can."

Apparently the Atomic Bomb was proposed at Bohemian Grove. "Historical records also clearly tell us that the Manhattan Project [Atomic Bomb] was planned, instituted and operated out of the Bohemian Grove," stated Hanson. Regarding the meeting at the Grove's river clubhouse on September 13-14, 1943, he wrote, "Bohemians make no effort to deny that this important historical moment took place at the Grove; in fact, they rather enjoy telling the story around the campfire."

He continued, "They publish it in their internal history books with a sense of pride. In an article entitled, *Gay Porn Star Services Bohemian Grove Members*, *The New York Post* reported on July 22, 2004, "Grovers privately boast that the Manhattan Project to develop the atom bomb was conceived on its grounds." *Extra! Magazine*, stated, "Policy speeches are regularly made by members and guests, and the club privately boasts that the Manhattan Project was conceived on its grounds."

"The atom bomb made this particular meeting at the Grove world famous," declared Hanson "but it was not an isolated case of business and government war planning through Bohemian Club facilities." Regarding the 1967 encampment, he wrote, "Nixon and Reagan sat down informally at the Grove to work a political deal wherein Reagan agreed to run only if Nixon faltered. Indeed Nixon's campaign was launched from the Bohemian Grove that year by means of his Lakeside Talk." In a 1989 Lakeside Talk, the plans for the Stealth B-2 Bomber were considered at the Grove, as well as plans for the European Union.

Hanson continued, "During the 2000 Grove summer camp, they announced Dick Cheney as George W. Bush's vice-presidential running mate from Bohemian Gove. *CNN* reported in July of 2000 that the decision had been partially made and that George Bush Jr. had been consulting with George Herbert Walker Bush ... at Bohemian Grove."

"But perhaps the most striking example of the elite Grovers' prior knowledge of future events," says Hanson, "is a speech given at the 1981 camp by the late nuclear scientist Dr. Edward Teller."

Hanson describes, "In this Lakeside Talk, he accurately predicted that the United States would go to war in the Persian Gulf—ten years prior to our first incursion into Iraq." Other Lakeside Talk topics have included, "World View," "The Recovered Memory Movement," "Agenda for America," "World Naval Power," "Rogues, Terrorists [and] ... National Security in the Next Century," "Spies and Terrorists: Confronting the Enemies Within," "Smart Weapons," and "Defining a New World Order."

One can only guess what the results of these talks have been. These speeches were delivered by Prime ministers, CEOs, scientists, chancellors, former directors of the CIA and FBI, and a variety of government officials.

Professor Phillips wrote, "On the surface, the Bohemian Grove is a private place where global and regional elites meet for fun and enjoyment. Behind the scene, however, it serves a very important function similar to 18th century French Monarchy scheming or the 19th century empire building of the British. The Bohemian Grove is an American version of race, gender and class elitism. It is the human process of building insider ties, consensual understandings, and lasting connections in the service of class solidarity. Ties reinforced at the Grove manifest themselves in global trade meetings, party politics, campaign financing, and top-down democracy. In a sense, they live in a self-made Bastille surrounded by power, prestige and privilege, and united in their fear of grassroots democracy." Hanson adds, "Conspiracy nuts think the Bohemian Club meets each summer to takeover the world. Get real. These guys ALREADY run the world."

"The Owl of Bohemia," is the club's symbol. Some members reportedly have their homes fraught with owls. "Club members collect owl statues, figurines, jewelry—anything with an owl on it," stated Hanson. He described, "You'll see them walking around the Grove wearing polo shirts embroidered with owls, owl belt buckles, owl bolo tiles, owl cufflinks, owl rings. ... The owl is found on all Bohemian materials from matchbook covers and doormats to the most elaborate Club publications." Apparently all of this merchandise can be purchased at the Bohemian Grove gift shop. The individual camps are decorated with wooden or stone owl statues according to Weiss. Even the 24-hour shuttle service within the grove is called, "The Owl" shuttle.

During the encampment, a 40-foot stone owl is worshiped in a ceremony called the *Cremation of Care*.(*3) The ceremony involves a mock-human sacrifice and is performed by men with torches dressed in red, black, brown and silver robes. Some of them have white skeletons painted on their faces. The ceremony takes place on an altar located at the base of the owl and lasts for about 15 minutes. Members admit to this ritual and information regarding it has appeared in mainstream publications. It was also filmed live, by Mike Hanson and Alex Jones. In fact, the Bohemian Club acknowledged this infiltration and even viewed the film. In a letter regarding the Cremation of Care, which was sent to a member of the Jones/Hanson team, they stated, “it’s about as innocent as anything could be.”

The New York Post acknowledged this, writing that members, “perform mock-druidic rituals that revolve around a 40-foot-tall stone owl. In one ritual, called ‘Cremation of Care,’ members wearing red-hooded robes cremate a coffin effigy of ‘Dull Care’ at the base of the owl altar.” “The annual gathering near the Russian River ... starts with the ‘Cremation of Care’ ritual,” echoed the *Sacramento Bee*, “in which the club's mascot is burned in effigy, symbolizing a freedom from care.” *The Daily Reveille* added, “The ceremony includes Grove members carrying torches and wearing ‘hooded robes.’” So, according to mainstream media, business leaders are performing mock human sacrifices at the Bohemian Grove.

Speaking from personal experience, Weiss wrote, “The cremation took place at 9:15 p.m. ... [and they] wore bright red, blue and orange hooded robes ... another priest or two appeared at the base of the main owl shrine, a 40-foot-tall, moss-covered statue of stone.” “Built to serve as a ceremonial site for traditional Bohemian rituals,” said Hanson, “the Owl of Bohemia is used yearly for the Cremation of Care Ceremony.” During the ceremony, Beethoven’s 7th symphony is played as well as funeral music.

Also, during the actual sacrifice, there is an auditory effect which is apparently intended to suggest that the sacrificial object is screaming in pain. The “Care” that these world leaders sacrifice, is obviously symbolic of the care or compassion that they must subvert in themselves, in order to carry out the acts necessary for world domination and the infliction of suffering that accompanies it. In a radio report entitled, *Exposé of the Bohemian Grove*, Professor Texe Marrs, a bestselling author, says that the ceremony would be “silly if it weren’t so evil.”

Some believe that this owl represents *Moloch*, or *Molech*, which is a deity in the Old Testament that children were sacrificed to. Hanson described, “There is no shortage of evidence that the rituals conducted at Bohemian Grove are derived from the rituals of ancient Babylon, Sumer, Canaan, Phoenicia, and Carthage."

He continued, "All the evidence seems to be pointing towards the Bohemian Grove's ritual being a contemporary manifestation of the ancient Cult of Isis. Among many other titles, Ishtar was known as 'The Lightbringer,' 'Light of the World,' 'Exalted Light of Heaven,' and 'Torch of Heaven and Earth.' Ishtar is also known as Venus, and many theologians see Ishtar as the female counterpart of Lucifer, who is also referred to as 'the Lightbringer.'" He concludes, "In the final analysis, the Great Owl of Bohemia symbolized the Cult of Isis, by symbolizing all four if its gods: Ishtar, Lucifer, and by virtue of the associated flames representing abandonment to Lucifer ... Satan."

During the ceremony, Jones and Hanson were surrounded by men groaning with sexual delight, saying, "Do it, do it," and "burn him, burn him." Both Jones and Hanson were sickened by the ritual. There was a feeling of disgust they felt that even the documentary, which they would later produce, could not express. Some of us have been in a similar situation, perhaps in the presence of another, where we could literally feel the evil emanating from their presence, which caused us to feel disgusted.

Dr Peck described, "Revulsion is a powerful emotion that causes us to immediately want to avoid, to escape, the revolting presence. And that is exactly the most appropriate thing for a healthy person to do under ordinary circumstances when confronted with an evil presence." "It's just really sick to be inside there," described Jones, and added, "words can't describe the evil that was in there." Jones sums up his description of the Bohemian Grove as, "a nightmarish, hellish place that I would expect Satan to reside in." It is a, "sick" place where members "carry out a Satanic ritual," added Professor Marrs.

Like its interrelated Think Tanks, Bohemian Grove appears to be directed by an inner-core, which most members may not be aware of. There have been rumors circulating for years that pedophilia, murder, rape and real human sacrifices have taken place at Bohemian Grove. If it is an established fact that these mock human sacrifices take place, and mainstream media has acknowledged this, then how far of a leap is it for some of these inner-circle members to perform real sacrifices?

Hanson and Jones had some occult experts analyze their footage of the Cremation of Care, and they concluded that what they witnessed was an outer sanctum of a cult that exists without the knowledge of most members. "In more recent years," proclaimed Hanson, "troubling new information has come to light in the testimony of CIA Mind Control survivors involving sexual abuse they suffered at the Bohemian Grove."

Former Senator John DeCamp stated that one of his clients spoke of an area later identified as Bohemian Grove, where his client was forced to witness and engage in Satanic activity. DeCamp wrote, "Paul [Bonacci] was taken by Larry King [no relation to the media personality] and others to a wooded area in California—identified after publication [of *The Franklin Cover-up*] as Bohemian Grove. There Paul and another boy were forced to ... [participate in unspeakable acts]." Apparently Boystown was used as a recruitment ground for young boys to be flown to Bohemian Grove and Washington to service wealthy pedophiles in positions of political influence.

Referring to this matter Professor Marrs said that these wealthy individuals were caught, "picking them up and flying them out to the Bohemian Grove where these orphans, evidently, were servicing these monsters." The professor continued, "He [Bonacci] was taken to ... the Bohemian Grove in California," and was "forced to do sex acts with other children in front of these wealthy perverts." DeCamp attested that, a snuff pornography film was made of these events. And Bonacci stated that the "men with the hoods" disposed of a murdered child dropped out of a helicopter over a wooded area.

Hanson commented, "Paul Bonacci was an eyewitness to the ritual rape and murder of a young boy at the bohemian Grove in the summer of 1984 ... [while] being forced to participate (at gunpoint) in a homosexual snuff film..." He added, "No teenage boy could possibly know these intimate details about the Grove unless he had actually been there. Even more, Bonacci had never even heard of the Bohemian Grove at the time he was taken there to 'entertain' the Bohos years ago."

Professor Marrs agreed, "He [Paul Bonacci] was forced to witness and participate in Satanic cult murder." He added, "Now think about this, why would he even know about such a place?" Interestingly, homosexual activity is known to take place at Bohemian Grove. Male and female prostitutes are often hired during the encampment.

"In the fall of 1992," proclaimed Hanson, "Paul was shown a black and white photo of the Grove's moss covered owl statue and quickly identified it as the site of the July 1984 snuff film described in DeCamp's book. This is not a game or the fanciful imaginings of a child. This is the cold-blooded ritual murder of a human being!" DeCamp added, "Understand that I didn't know that this place he describes was Bohemian Grove back then, nor did the kid when he was writing it."

Speaking of just one of the many sudden deaths that occurred surrounding this investigation, Jones stated, "The senator was hired by the republicans to disprove these stories, [but] he found out it was true. Then his investigator had his plane blow up, right after he called him [DeCamp] and said I'm flying out to Chicago ... I've got the goods on these people." The investigator and his son were killed.

Another person with first-hand experience of grove activities is Cathy O'Brien. Hanson comments on O'Brien's account of the grove, writing, "O'Brien writes of her visits to the Bohemian Grove, [which were] secretly approved by top insiders. Her programmed mission was to take part in porn films shot at the Grove, and to entertain in the various secret live sex rooms located within the massive property." She says she was often prostituted to various government leaders which was covertly filmed for blackmail purposes.

In her documented autobiography, *Trance Formation of America*, former government mind-control victim Cathy O'Brien writes, "Bohemian Grove is reportedly intended to be used recreationally, providing a supposedly secure environment for politically affluent individuals to 'party' without restraint." O'Brien was apparently used by the CIA at the Bohemian Grove to compromise politicians. "My knowledge of these cameras," she describes, "was due to the strategically compromising positions of the political perpetrators I was prostituted to in the various kinky theme rooms." Apparently, for the inner-core members, the grove is equipped with several, "perversion theme rooms," which include necrophilia, group orgy rooms, and others too graphic to mention.

She continues, "Slaves of advancing age or with failing programming were sacrificially murdered 'at random' in the wooded grounds of Bohemian Grove." "Rituals were held at a giant, concrete owl monument," where "I witnessed the sacrificial death of a young, dark-haired victim." She adds, "no memory ... is as horrifying as the conversations overheard ... pertaining to implementing the New World Order." "The solution being debated," she describes, "was not pollution/population control, but mass genocide of 'selected undesirables.'"(*4)

"Even more difficult to ignore," wrote Hanson, "is the fact that Bonacci's tale jives with that of Cathy O'Brien, who was forced to participate in a similar snuff film where she witnessed the murder of another young woman at the Grove. Paul Bonacci and Cathy O'Brien have never met ... yet they tell nearly the same story" but "no official investigation has been done..." The grove consists of over 2,700 acres of redwood forest. It would not be difficult to conceal an actual murder. Hanson wrote, "A lot of the people who work at the Grove don't even get to go over to where the Owl is. That's a whole separate part from where they work and they are confined to a certain area. We talked to some Grove employees who don't even know that the owl statue is there."

For the most part, it's common knowledge that human sacrifices happened in ancient times. These sacrifices were apparently conducted by the financial elites of those periods. Even if these are only mock human sacrifices, at the very least, the people involved are still performing the rituals symbolic of the subversion of their conscience. Dr. Peck noted, "Those who are evil" exhibit a "brand of narcissism so total that they seem to lack, in whole or in part, this capacity for empathy."

So, if one of the major steps on the path to evil is the removal of compassion, or care. And if it is an established fact that world leaders are emulating this removal of care for humanity by sacrificing it, then this is a serious problem. However, I tend to believe that much worse things go on at the grove, because I've concluded that the people who run this planet are definitely monsters.

Other than engaging in occult rituals and planning policy without the public's approval, what's the purpose of the grove? O'Brien writes, "My perception is that the Bohemian Grove serves those ushering in the New World Order."

Former FBI Senior Special Agent Ted L. Gunderson has 27 years experience investigating Satanic cults. In his report, *Techniques Used To Silence Critics*, he sums up the Bohemian Grove as a "Satanic Cult" which is "actively practicing behind the scenes ... to insure the passage of regressive legislation that takes away many of our Constitutional rights contained in the Bill of Rights." He says this legislation includes, "Homeland Security and the USA Patriot Act." Gunderson's claim that these major decisions were made at the grove is consistent with what has already been admitted in mainstream news. He also concludes that these Satanists control all of the major Federal Government regulatory agencies.

Run by Satanists

According to this information, the planet is essentially controlled by wealthy Satanists. Author Perloff wrote, "Many centuries ago, the Hebrew Old Testament ... and the Christian New Testament ... warned of an evil, one-world government." He explained "Many notables of the American Establishment have given themselves over to one side in this conflict, and it is not the side the ancient scriptures recommend."

On her *Road to Freedom Show*, which covered the topic of Satanism and organized stalking, Eleanor White proclaimed that the upper level of this international stalking group is Satanic in nature. "In my view," she says, "nothing else could account for such merciless destruction of innocent peoples' lives." Now that we understand that Satanic cults composed of world leaders exist, this becomes perfectly logical.

Unsurprisingly, the FBI's *Project Megiddo* "educated," law enforcement, by warning them of paranoid extremists, who believe they are being persecuted by the Satanic U.S. government. "Almost uniformly, the belief among right-wing religious extremists is that the federal government is an arm of Satan," they say. "By extension," they add, "the FBI is viewed as acting on Satan's behalf."

"I proudly served in the FBI for more than 27 years," countered Gunderson, "and realize that there are thousands of loyal, honest and dedicated citizens in all branches of the government, including the FBI, the CIA and other intelligence agencies. However, it is apparent that upper level management in the FBI and CIA and other intelligence agencies are under the control of ... satanists." Gunderson says there is an active disinformation program launched by the U.S. Government to discredit those who speak out against these wealthy Satanists.

Summary

People who essentially run the planet, have openly declared that they engage in occult rituals, which multiple reliable sources conclude to be Satanic in nature. These rituals include the sacrifice of compassion, and oaths of obedience which are pledged to a cause. The evidence suggests that these rituals are symbolic pledges to carry out a political agenda known as the *New World Order*.

Elite Satanic cults, the drug trafficking trade, and the international child sex-slave business are interconnected, according to these researchers. To this I would add, the Hidden Evil. These researchers contend that this is a multi-billion-dollar worldwide enterprise. The fragments that occasionally appear in mainstream media are merely a tiny fraction of an undercurrent of interconnected programs, which are part of Satanism. The occasional media pieces are diffused by federal law enforcement which botch investigations, and by official "experts" who assist with cover-ups by discrediting witnesses.

The documentation presented portrays a heavy cross membership between the elite Think Tanks and Satanic cults. All of these groups work in concert to essentially control the planet. The evidence provided also indicates that this Satanic thread runs from the top-down, and encompasses a network of highly organized crime. This influence runs from the supra-governmental organizations down into the executive branch, and federal level. From there, it extends down to state and local governments where it manifests as the Hidden Evil.

Citizens on the lower level, who participate in stalking targeted people, are mostly not aware that their seemingly helpful group is one small component of a larger, very malicious structure. If they were to follow this command structure upward, they would most assuredly find that it meets at a hub, where other destructive activities originate. The Hidden Evil exists as an offshoot alongside other activities at the highest levels, emanating basically, from Satanism.

Endnotes

* See *Shadows of Power*, by James Perloff.

*1 See *Americas Secret Establishment*, by Professor Antony Sutton, *Dark Majesty*, by Professor Texe Marrs, and *Bohemian Grove,* by Mike Hanson.

*2 Hanson says 1873, Peter Phillips and Philip Weiss say 1872.

*3 The owl is 40, or 50 feet tall and made of either stone or concrete. No non-member has ever measured it. Jones says it could be 45-50 feet, Hanson says it could be 45, the *New York Post* and Weiss say 40 feet.

*4 Cathy's experience leads her to conclude that those implementing the New World Order are drawn from the CFR and other elitist groups. Interestingly, in her second book, *Access Denied: For Reasons of National Security*, she wrote about the frequent use of, what she later discovered to be NLP, by her government handlers.

Chapter 26

Techniques to Discredit

Sutton and Wood warned, "Anyone in the U.S. who promotes unwelcome news for the elite receives some unwelcome attention in return." This attention includes the use of experts, mainstream media, and anti-hate organizations, for well-coordinated character assassination.(*) According to author Perloff, a document released by the House Committee on Un-American Activities, on September 23, 1956 entitled, *Propaganda and the Alert Citizen* outlined the tactics reportedly used by the Establishment to discredit opposition.

Quoting from this document, Perloff wrote, "The tactic used had a prototype in a directive issued by the Communist Party... It read: When certain obstructionists become too irritating, label them, after suitable build-ups, as Fascist or Nazi or anti-Semitic, and use the prestige of anti-Fascist and tolerance organizations to discredit them. ... In the public mind constantly associate those who oppose us with those names which already have a bad smell."

"Because the Establishment controls the media," wrote Allen, "anyone exposing the Insiders will be the recipient of a continuous fusillade of [criticism] ... from newspapers, magazines, TV and radio. In this manner one is threatened with the loss of 'social respectability' if he dares broach the idea that there is organization behind any of the problems currently wrecking America."

"Smear tactics" from "left and right" are "standard operating procedure," explained Sutton and Wood. Even if you don't believe in the phony *left/right* political spectrum and have no political affiliation, you may be assigned one. Then the financial elite can use the opposite political party expressed in the mainstream news for character assassination. These labels, describe Sutton and Wood, "divert attention from responsible reasoned criticism with no attachment to a synthetic political spectrum."

"If you assemble the evidence, carefully present your proofs, and try to expose these power seekers, the Establishment's mass media will accuse you of being a dangerous paranoid who is—dividing—our people," Allen said. In his book, *With No Apologies*, Senator Goldwater describes being on the receiving end of the Establishment's "professional scandalmongers." Apparently, because he was not controlled by the Establishment, the *New York Times* and *New York Herald Tribune* were used to discredit him during his campaigns.

Perloff wrote, "Barry Goldwater was the Republican nominee and, as such, was the first GOP Presidential candidate in decades it [the CFR] had not controlled. Indeed, Goldwater represented nearly everything the Establishment was against. For that reason, the mass media was arrayed against him, and he was falsely characterized as a fanatic who would start a nuclear war and snatch social security checks from the elderly."

Investigations into Crimes and Practices of the Financial Elite are Thwarted—Witnesses Discredited

DeCamp explained that during his investigation of a child prostitution ring involving Satanism which lead to Washington D.C., that "Articles began to appear in the *Omaha World-Herald*, aimed to discredit the witnesses and intimidate any other potential child victim-witnesses from testifying." He cites some examples including the *World-Herald's* charge that it was a "hoax based on rumor." Editorial attacks ensued with titles such as, *Grand Jury Did Its Job; The Insults Are Intolerable*, and *Schmit Panel Can't Duck its Responsibility in Hoax*.

The front page of the *Omaha World-Herald* on July 25, 1990 ran the headline, *Grand Jury Says Abuse Stories Were a Carefully Crafted Hoax.* And an article entitled, *Former Legislator's Angry Memo Turns Sober Nebraska on its Ear*, which appeared in the March 17, 1990 edition of the *Kansas City Star*, stated that the memo was, spawning swirls of gossip. The same article described that the accusations against these prominent people was like "insulting God." Compare these attacks to the criticism General Butler received by the mainstream news after exposing the plot by Wall Street to overthrow the U.S. Government, or the harassment the Reece Committee received when investigating the Tax-exempt Foundations.

One of the most damaging dirty tricks used against DeCamp was an *anonymous complaint* from a concerned citizen, which was logged with the Department of Social Services, stating that he and his wife were abusing their daughter. This tactic was apparently used to stop his investigation and knock him out of the senate race. The cover-up of a similar event occurred in Minnesota, where prosecutor Kathleen Morris came under savage attack from the media when investigating a child sex ring that was apparently linked to the financial elite. After experiencing considerable pressure, Morris relinquished the investigation to Attorney General Hubert Humphrey Jr. Humphrey soon closed the case due to insufficient evidence.

When the elite are linked to these types of crimes, evidence become missing, leads are not followed, people are discredited, jailed, disappear, or have accidents. These occasional leaks are part of a steady stream of corruption that only occasionally surfaces in the mainstream news. The discrediting and harassment tactics used during these cover-ups can happen anywhere the elite need to eliminate exposure to their activities. DeCamp calls it an "international organized crime syndicate, engaged in pedophilia, pornography, satanism, drugs, and money-laundering," which is "protected ... by federal authorities."

William H. Kennedy describes an organized kidnapping ring in Belgium linked to the financial elite, in his book *Satanic Crime: A Threat in the New Millennium*. Using mainstream news, Kennedy explains that during these investigations, the police were consistently incompetent, made frequent mistakes, and failed to follow leads. These seemingly deliberate acts resulted in further deaths and kidnappings. The presiding judge, Jean-Marc Connerotte, secured the arrest of suspect Marc Dutroux. Dutroux escaped from prison once and was poised for a second escape before he was found with a handcuff key.

According to Kennedy, Dutroux provided kidnapped girls to a Satanic cult which included members of the Bilderbergers. Judge Connerotte required a bulletproof vehicle and armed guards after learning that a contract had been taken out on his life. He was eventually dismissed from the bench. The removal of Connerotte, and the cover-up, triggered a peace march by 300,000 outraged Belgians who demanded reform of the political and judicial system.

The Hidden Evil is Beyond a Congressional Investigation

Similar tactics were used during the investigation of the Tax-exempt Foundations and their Think Tank interlocks during the Reece Committee hearings. Wormser described the violence of the attacks against the committee as "amazing." The Reece Committee itself observed that because the Tax-exempt Foundations and Think Tanks control the mainstream news and the government, that they were unable to conduct their investigation.

The committee stated specifically, "The far-reaching power of the large foundations and of the interlock has so influenced the press, the radio, and even the government that it has become extremely difficult for objective criticism of foundation practices to get into news channels without having first been distorted, slanted, discredited, and at times ridiculed."

In his book, *How the World Really Works*, Alan B. Jones considers the account of another Reece Committee insider, Norman Dodd, the Research Director. Dodd attests that death threats, harassment of witnesses, stalking, surveillance, blackmail, and framings were the real reasons the trials were discontinued.

The Reece Committee declared, "The pressure against Congressional investigation has been almost incredible. As indicated by their arrogance in dealing with this Committee, the major foundations and their associated intermediary organizations have intrenched (sic) themselves behind a totality of power which presumes to place them beyond serious criticism and attack." Smoot commented, "As we have seen, two different committees of Congress—one Democrat-controlled and one Republican-controlled—have tried to investigate the ... tax-exempt foundations which are interlocked with, and controlled by, and provide the primary source of revenue for, the Council on Foreign Relations and its affiliates."

Smoot continues, "Both committees were gutted with ridicule and vicious denunciation ... by internationalists in the Congress, by spokesmen for the executive branch of government, and by big respected publishing and broadcasting firms which are a part of the controlled propaganda network of the Council on Foreign Relations." So according to the results of a congressional investigation, the Tax-exempt Foundations and Think Tanks are beyond investigation.

There was another attempt to investigate the Think Tanks by Congressman Larry McDonald. "In 1980, the American Legion national convention passed Resolution 773, which called for a congressional investigation of the Trilateral Commission and its predecessor, the Council on Foreign Relations," wrote author Marrs. "The following year," proclaimed Marrs "a similar resolution was approved by the veterans of Foreign Wars (VFW)."

"Congressman Larry McDonald," continued Marrs, "introduced these resolutions in the House of Representatives but nothing came of it." Unfortunately Congressman McDonald, who opposed the Establishment, died in a curious incident involving a Russian missile which blew up Korean Airlines 007 on September 1, 1983. The bottom line is simply this: The Establishment will not allow the various organizations which it controls to stop the Hidden Evil.

According to these researchers, some with first-hand experience, anyone who exposes the activities of the elite usually receives an assortment of character assassination from the mainstream media, *anti-hate* organizations, and official *experts*. Lawsuits may be filed. An IRS audit may occur. Other tactics are used, such as being threatened, framed, blackmailed, and murdered. Accidents which result in death or injury may be reverse engineered to appear as though they were not planned.(**)

Many of these attacks are tactics within the Hidden Evil. According to the HUAC document, some people are allowed to gain enough momentum until they are in pubic view so they may be smeared by the mainstream media or other organizations under elite control. By this account, it is probably easier for the elite to ruin famous people, because they require no "suitable build-ups." Anyone who falls into disfavor with the elite can be ruined by these methods.

"The money powers prey upon the nation in times of peace and conspire against it in times of adversity. It is more despotic than a monarchy, more insolent than autocracy, and more selfish than bureaucracy. It denounces as public enemies, all who question its methods or throw light upon its crimes."
-Abraham Lincoln

The Dangers of a Congressional Investigation

Congress, in general, is under the absolute control of this mafia. As previously demonstrated, this has been accomplished through years of infiltration, blackmail, intimidation and bribery. The elite, who own the United States, literally, will never allow an investigation into the Hidden Evil. They will never allow the system which they control to stop this.

A congressional investigation may be done for the deliberate purpose of convincing the general population that it does not exist, and to publicly discredit those speaking about it by portraying them as a few paranoid conspiracy theorists. Any investigation into the Hidden Evil by congress or another organization under the control of the Establishment will result in a *whitewash.*

These whitewashes by congress have unfolded repeatedly. Consider that in addition to the outside attacks, the Reece Committee was also killed from within. This included receiving "palpably inadequate" funding, a reduction in research staff, denial of evidence, and general stonewalling. "Mr. Reece understood, soon after our investigation started, if not before, that we would be met with every obstacle which could be put in our way," stated Wormser.

Other congressional whitewashes include the Roberts Commission (Pearl Harbor cover-up), and the Kerry Subcommittee (Iran Contra Scandal, Federal Drug Trafficking). The Warren Commission (investigating the death of JFK) is said to have been heavily whitewashed, the 9/11 Commission was a whitewash, and the McCormack-Dickstein Commission was partly a whitewash.(***) Some of these commissions were led by members of the CFR.

The United Nations

If the UN investigated the Hidden Evil it would be a maneuver to whitewash and discredit targets. According to some researchers, the United Nations is a creation of the CFR. The UN would like us to believe that they're an organization that acts as a non-partisan referee, which reluctantly intervenes in world conflicts for the betterment of all parties concerned. Apparently, the idea is that if enough chaos is caused and the UN is portrayed in a positive manner, then the masses will believe that a world government under the control of the UN is the only hope for world peace.

In a chapter entitled, *United Nations and World Government Propaganda*, Smoot commented, "All American advocates of supra-national government, or world government, claim their principal motive is to achieve world peace. Yet," he says, "these are generally the same Americans whose eager interventionism helped push America into two world wars."

"By 1945, the Rockefellers were ready," wrote Allen. "Grandson Nelson was one of the 74 CFR members at the founding meeting of the United Nations in San Francisco." Perloff added, "Of the American delegates at the founding UN conference in San Francisco, more than forty belonged to the CFR." "Nelson and his brothers donated the land for the United Nations complex along the East River in New York," wrote Allen, and added, "possibly because they did not want the new headquarters of their World Government to be more than a short taxi-ride away from their penthouses."

Author Marrs portrays the UN as, "an outgrowth of the old League of Nations," which he describes as a failed attempt at, "world government." "A primary mover of this and subsequent actions to establish a United Nations was John Foster Dulles, who had helped found the Council on Foreign Relations," wrote Marrs.

Marrs continued, “Considering Dulles and the other CFR members [are] behind the creation of the UN, it is no surprise to find that organization today supervising the International Bank for Reconstruction and Development (commonly called the World Bank) and the International Monetary Fund (IMF).” Author Alan B. Jones sums up the UN as a front for the banking elites to be used as the core of a political structure for a world government, using the IMF as the world central bank.

“In 1948, the State Department created the U.S. Committee for the UN ... as a semi-official organization to propagandize for the UN,” conveyed Smoot. The CFR dominates this committee, he says. He also contends that the UN Charter, which the CFR would like to have the US Constitution replaced with, is essentially a charter for *enslavement*.

“The ‘world peace’ aspects of the United Nations were emphasized to enlist support of the American public,” wrote Smoot. However, he adds, “the UN Charter really creates a worldwide social, cultural, economic, educational, and political alliance—and commits each member nation to a program of total socialism.”

Project Megiddo

Project Megiddo is described by the FBI as a, “strategic assessment of the potential for domestic terrorism in the United States...” The document was apparently circulated to law enforcement throughout the country in 1999 to help identify domestic terrorist activities spawned by the year 2000 craze. It basically says that anyone who holds, “political beliefs relating to the New World Order (NWO) conspiracy theory,” is “paranoid.” “The challenge to law enforcement,” it states, “is to understand these extremist theories.”

And just what are these paranoid conspiracy theories that the FBI is “educating” law enforcement on? According to the FBI, “The NWO conspiracy theory holds that the United Nations (UN) will lead a military coup against the nations of the world to form a socialist or One World Government.” The FBI lists other groups that a paranoid conspiracy theorist might say are part of this socialist overthrow as: the Council on Foreign Relations, the Bilderbergers and the Trilateral Commission. They add, “Law enforcement officials will probably notice different versions of this theory, depending upon the source.”

They continue, “Law enforcement officers, as well as military personnel, should be aware that the nation's armed forces have been the subject of a great deal of rumor and paranoia circulating among many militia groups." They describe these militias as "Patriots," and "right-wing" groups who, "believe they are being persecuted by the satanic government of the United States."

It looks like the CFR-controlled FBI laid the groundwork to portray anyone telling the truth as a dangerous paranoid. Megiddo is clearly CFR *Damage Control.* It seems to be an employment of the tactics contained in the HUAC document that Perloff warned about. Remember? "In the public mind constantly associate those who oppose us with those names which already have a bad smell."

Continuing, they stated, "Use of this term [New World Order] within militia circles became more common after President Bush starting using it to refer to ... using international organizations to assist in governing international relations," and added, "The term One World Government is also used as a synonym for the New World Order."

According to a document written by Dennis L. Cuddy, PhD, entitled, *Chronological History of the New World Order*, the term has been used since the mid 30s by a variety of people including multiple members of the CFR such as Nelson Rockefeller, Henry Kissinger, and Richard Nixon. It has been mentioned in CFR's publication *Foreign Affairs*. H. G. Wells spoke of it, as did Mikhail Gorbachev. It has been used by George H. W. Bush on multiple occasions.

"The term 'New World Order,'" says Cuddy, "has been used thousands of times ... by proponents in high places of federalized world government." In the last several years it has been used by leaders of a variety of nations who are obviously under the control of the financial elite.

So according to Project Megiddo, the New World Order does exist and is beneficial. However, when we consider the publications of the "international organizations" that are assisting with "governing international relations" that Bush has been a member of, such as the CFR, the TC, and the Bilderbergers, it becomes clear that their intentions are to reduce civil liberties and merge America into a single world government controlled by them. It's obvious to me that Megiddo is disinformation, used for the purpose of discrediting anyone who is attempting to warn law enforcement about the fate of America.

Speaking of Project Megiddo's predecessor, Former FBI Senior Special Agent Ted L. Gunderson wrote, "In 1996 the FBI Phoenix Division issued a pamphlet to local and state law enforcement agencies that set forth the profile of various terrorists in the United States. ... This profile includes: Defenders of the U.S. constitution against the federal government and the UN (Super Patriots) [and] ... anyone who makes references to the United States Constitution..." He warns, "An active Federal Government Disinformation Program ... was instituted for the purpose of discrediting ... [people] who report the truth about the government."

Apparently, according to Gunderson, federal agencies release these documents as *disinformation* for the sole purpose of discrediting people who begin to wake up and tell the truth. I wonder—if we know that these two documents exist, then is it possible that more damaging information has been given to law enforcement to discredit the truth movement?

"Distrust of the government by not thousands but tens of millions of U.S. citizens is confirmed in public opinion surveys," stated Dr. Alexander in his book, *Future War*. "Many of these conspiracy theory adherents believe that the government—or some other supranational organization—is attempting to take freedom away from citizens." As expected, he lists these supranational organizations as, the Trilateral Commission, the Bilderbergers, and the Council on Foreign Relations, and The Order of Free and Accepted Masons.

He continues, "Some of them see non-lethal weapons as tools to facilitate those objectives. They believe that these weapons potentially could be used to enslave them..." "Such systems," he says, "are usually rumored to be designed and controlled by some unspecified 'They' who are the ultimate powerbrokers." Evidence has already been provided which undeniably proves that the Trilateral Commission, and the Council on Foreign Relations have infiltrated the executive branch. I wonder how many of those tens of millions of conspiracy theorists are targeted but don't know it.

False Memory Syndrome Foundation

According to some researchers, one example of an organization that was launched for the sole purpose of discrediting people is the False Memory Syndrome Foundation (FMSF). In his book, *Mass Control: Engineering Human Consciousness*, author Jim Keith refers to the FMSF as a "pseudo-scholarly" front organization, consisting of "a group of psychiatrists whose mission is to prove that cult abuse and mind control are figments of the imagination."

According to Keith, many of the psychiatrists on the FMSF board have been linked to mind-control experiments and the military. They are called in to court cases to discredit ritual abuse survivors.

Author Constantine's analysis is similar, he proclaims that the FMSF consists of a panel of experts which are used to discredit therapists and victims that come forward with cases of Satanic Ritual Abuse (SRA) and experimentation.

Professor Jennifer J. Freyd explains in her book, *Betrayal Trauma*, how the FMSF, her colleagues, and the mainstream media attempted to discredit her research into repressed memories. Her mother, Pamela Freyd, a founding member of the FMSF, also participated in the attacks against her credibility.

In their book, *Trance Formation of American*, Cathy O'Brien and Mark Phillips arrive at the same conclusion. These and other authors contend that the CIA found it necessary to create this professional organization in order to discredit abuse claims.

Mainstream News is Used to Discredit Targets

Anytime mainstream media addresses an issue like this there is a very real danger of it being an attack. This was clearly the case with *The Washington Post's* January 14, 2007 article, entitled, *Mind Games*, or more appropriately, it should have been called, *Word Games*.

Catherine Graham, the owner of *The Washington Post*, is a member of the Council on Foreign Relations, the Trilateral Commission, and the Bilderbergers. The *Post* and most other mainstream news are controlled by the very elite Think Tanks that I'm alleging are responsible for this covert war against the civilian population.

While I can't say for sure how this story was arranged, my guess is that representatives from the *Post* approached some TIs and portrayed themselves to be genuinely interested in covering the issue. Naturally, TIs anxious to get the word out, cooperated. But what resulted was a standard discrediting article. Although it did explain that thousands of people in the NATO countries claim to be targeted, it basically described them as *delusional*.

Some common tactics to discredit are to portray facts as beliefs, to ignore evidence, and portray people as a haters or paranoids. This article has done all three. They realize that most of their readers won't research this story for themselves. Also, no websites devoted to this phenomenon were listed so their readers could follow-up.

"The idea of a group of people convinced they are targeted by weapons that can invade their minds has become a cultural joke," they wrote. Referring to targeted people they stated, "In their esoteric lexicon, 'gang stalking' refers to the belief that they are being followed and harassed: by neighbors, strangers or colleagues who are agents for the government." Notice how they used the word, "belief."

The article continued, "Gloria Naylor, a renowned African American writer, seems to defy many of the stereotypes of someone who believes in mind control." Once again, playing games with words. In this case implying that mind-control is just a belief and that projects such as MKULTRA did not take place. This is not a belief, it's a fact. Later in the article they acknowledged MKULTRA but portrayed it as simply an, "infamous CIA program that involved, in part, slipping LSD to unsuspecting victims." No mention was made of the other "parts," such as the children who were sexually abused, or of the other people who were brutally tortured and had their lives destroyed. This filtering out of information is basically, lying.

They described one TI, probably an informant, who "criticizes what he calls the 'wacky claims' of TIs who blame various government agencies or groups of people without any proof." Of course this torturous program has not been *proven* in a court of law. And for reasons previously mentioned—it never will be. But there is plenty of *evidence* that it exists.

The article went on to describe the TI mentioned above (a doctor), who was "hoping to prove that V2K, the technology to send voices into people's heads, is real." Once again, playing games with words. As previously demonstrated, mainstream publications announced that V2K was possible in the 70s. Once more, turning a fact into an opinion or belief. They are fully aware that most people either don't have the time or the inclination to do the research.

Of course no hit story would be complete without the Establishment's use of official "experts" who assist them in damaging potential witnesses. Regarding MKULTRA, the article quoted from their interview with Dr. John B. Alexander, one of the alleged creators of this program. They wrote, Alexander acknowledged that "there were some abuses that took place," but added that, on the whole, "I would argue we threw the baby out with the bath water."

More games with words. First, dismissing MKULTRA as just "some abuses," leads those who have not studied it (most of their readers) to conclude that there were a few slip-ups in an otherwise beneficial program. Second, implying that it stopped. Funny, Dr. Alexander must know that it continues in its advance form. As I've demonstrated, he and his colleagues helped create these weapons on behalf of their CFR handlers, for the "New World Order" which they've trumpeted in their articles.

The *Post* continues citing more "experts." "Ralph Hoffman, a professor of psychiatry at Yale who has studied auditory hallucinations, regularly sees people who believe the voices are a part of government harassment," they said. "Not all people who hear voices are schizophrenic, he says, noting that people can hear voices episodically in highly emotional states. ... People who think the voices are caused by some external force are rarely dissuaded from their delusional belief, he says."

Quoting Professor Hoffman, they wrote, "These are highly emotional and gripping experiences that are so compelling for them that ordinary reality seems bland." There are several assumptions that most readers of this statement will come away with. First, that V2K doesn't exist. Second, that all people complaining that an available technology is being used to harass them are definitely delusional. Third, that the government has not targeted the civilian population with advanced technology in the past.

Continuing their assault, they stated, "Scott Temple, a professor of psychiatry at Penn State University who has been involved in two recent studies of auditory hallucinations, notes that those who suffer such hallucinations frequently lack insight into their illness. Even among those who do understand they are sick, 'that awareness comes and goes... People feel overwhelmed, and the delusional interpretations return.'"

So, Professor Temple studied auditory hallucinations, but did he study a technology that can cause them? The article doesn't say. Once more, the implication is that it doesn't exist and that all of those complaining about it are delusional.

Remember, doctors are highly complicit in the cover-up of this program. In addition, as previously demonstrated, the elite-owned Tax-exempt Complex provides the grants to these professors to do research projects with *pre-determined* outcomes. In other words, when the Establishment wants an official "expert" to prove a point, they'll provide them with grants to carry out a project which *they've* created. These college professors are often *owned* by the Establishment.

"The very 'realness' of the voices is the issue—how do you disbelieve something you perceive as real? That's precisely what Hoffman, the Yale psychiatrist, points out: So lucid are the voices that the sufferers—regardless of their educational level or self-awareness—are unable to see them as anything but real. 'One thing I can assure you' Hoffman says, 'is that for them, it feels real.'" The people at the *Post* accomplished their obvious goal. My guess is that this story ran because information was beginning to leak out to the public.

Summary

Those who expose the financial elite typically receive character assassination from radio, TV, print media, official "experts," and "anti-hate" organizations. A combination of framings, murders, "accidents," and lawsuits may occur. "Educational" material is distributed to law enforcement to help them understand the "paranoid extremists" who believe that multinational corporations are installing a global dictatorship.

Endnotes

* DeCamp stated that the elite used the mainstream news, the Anti Defamation League (ADL), and the Cult Awareness Network (CAN) to discredit him and other witnesses. The Establishment also used the mainstream news and the ADL to discredit the Reece Committee. For examples of this see *Foundations: Their Power and Influence*, by Rene Wormser. Also see *The Invisible Government* by Dan Smoot who describes how the Establishment labeled people anti-semitic and racists who investigated the Tax-exempt Foundations under the Cox Committee.

** DeCamp spoke of learning about an impending lawsuit against him which was to be used to destroy him in public view. There were 15 suspicious deaths of potential witnesses surrounding the Franklin case. One included chief investigator Gary Cordori who conveyed to DeCamp that he had evidence that would "blow the story wide open." Senator Loran Schmit who participated in the investigation received death threats and attempts to ruin him financially. DeCamp also stated that during the investigation multiple witnesses received IRS audits. Sutton and Wood contend that IRS audits are often used to intimidate critics of the Establishment. During Cointelpro the IRS was used for harassment, as described in the book, *War At Home*, by attorney Brian Glick. See also *The Autobiography of Martin Luther King Jr.* for how he was arrested for falsifying his tax returns in retaliation for his nonviolent direct action to affect social change.

*** For the Iran Contra whitewash see the book, *Powderburns* by Celerino Castillo (former DEA) and Dave Harmon. Also see *Cocaine Politics* by Peter Dale Scott and Jonathan Marshall. For the Reece Committee whitewash see the books, *The Invisible Government* by former FBI Agent Dan Smoot, *How The World Really Works*, by Alan B. Jones, and *Foundations: Their Power and Influence*, by Rene Wormser. For the Pearl Harbor cover-up (The Roberts Commission) see *Rule by Secrecy*, by Jim Marrs and *The Shadows of Power* by James Perloff.

Chapter 27

Why it Remains

This program is thoroughly covered-up by mainstream institutions. Its designers had enough global influence to block off all standard escape routes. This is includes the medical system, the legal system, and worldwide organizations which claim to exist to help victims.

McKinney's investigation into the claims of this harassment revealed a general "Failure of 'Establishment' Support Systems." This makes sense when one understands that the Hidden Evil is the Establishment's policy. She described that those who tried to resolve their issues through existing support mechanisms encountered: Apathy and indifference on the part of congress and state legislators, attempted discrediting by psychiatrists, lack of interest and/or competency in legal circles, little or no assistance from police, and a refusal of both the ACLU and Amnesty International to intervene.

Apathy and Indifference

Congress and state legislators have been contacted by *many* people. They know this program exists and are allowing it to continue. They are unwilling or unable to stop it. As already demonstrated, congress is subservient to the Think Tanks. "They just won't intervene," said McKinney, regarding efforts to contact state representatives about this program.

Discrediting by Psychiatrists

Some targets that have reported this program to psychiatrists have been labeled mentally ill. This tactic was heavily used in Russia, where they would subject people to the penal and mental health systems who were critical of the dictatorship. Professor Marrs observed, "The communists of the USSR were long engaged in forcing their 'malcontents'—those who oppose the monstrous authoritarian system—to be 'treated' in crowed psychiatric hospitals." He continued, "There, confined to a secluded prison ward, they were treated for 'mental illness' by becoming injected with massive does of mind ... destroying drugs."

"If you go to a medical doctor you do not talk about [it]," warned McKinney, "because many are also involved." Some doctors who were aware of existing conditions of targeted patients deliberately put them on a regimen of drugs in order to worsen pre-existing conditions. The CIA is apparently famous for using this tactic. In his book, *Journey Into Madness: The True Story of Secret CIA Mind Control and Medical Abuse*, Gordon Thomas wrote, "Nothing I had researched before could have prepared me for the dark reality of doctors who set out to deliberately destroy minds and bodies they were trained to heal."

He continued, "The realization that physicians are part of a killing machine provokes a special horror." Apparently the Stasi would use doctors to help destroy enemies of the state as well. Fulbrook wrote, "There have also been suspicions that symptoms of mental illness were actually created by 'medical' treatment."

Lack of Legal Assistance

This cannot be resolved through legal means. Attorneys will not generally assist. Some have allegedly deliberately sabotaged cases. In Germany, targets found themselves in a similar situation; Funder stated, "There was no room for a person to defend themselves against the State because all the defense lawyers and all the judges were part of it."

Police will not assist. Standard practice seems to be dismissing targeted individuals that report this as lunatics. Some police acknowledge the existence of this program, but won't intervene with what they believe to be operations by U.S. Intelligence. Federal law enforcement will not assist either.

Apparently, the FBI has been contacted by a great number of people who claim to be on the receiving end of V2K and other Non-lethal harassment, but dismiss them as, "mentally disturbed persons." McKinney stated, "Writing to the various [federal] agencies and calling and meeting with them serves no useful purpose either because they will say there are no laws prohibiting these types of activities."

The ACLU and Amnesty International Will Not Intervene

Even though both organizations acknowledge receiving many complaints, they have offered little or no assistance to targeted people. Obviously, the top leadership in both organizations has been told to stand down while terrorists and undesirables are dealt with. This excuse has been used countless times throughout history—"we need to temporarily suspend civil liberties for some people in order to protect the state."

What this means is that those in charge of the dictatorship are doing their roundup. This is standard dictatorship installation. *National Security* is very likely the reason used to keep these organizations from assisting targeted people. This program thrives so the elite can target anyone who they either dislike or think is a potential threat to their control.

The government can't control it, because those who created it control the government, and use government agencies as instruments of persecution. Compare this to the group stalking program used in Germany, where the definition of "enemy" increased as time went on. "The number of informal collaborators, or IMs (inoffizielle Mitarbeiter), exploded over the years," declared Fulbrook. Remember, the Stasi and their colossal network of citizen informants existed to keep the criminals who ran the state in control. According to *CNN*, this happened until about 1/3rd of the population was targeted in some manner.

Denial and Fear

This is a topic that people may not want to believe, even though it is true. Most people don't want to face the horror of acknowledging that something like this is going on. They don't want to entertain the possibility that, yes—it's happening again, their own government has turned into a monster. When you make someone aware of the possibility that their government has been hijacked and has sinister plans for the population they're presented with a choice. They can believe you and have their world turned upside down. Or they can deny that it exists. Some people don't have the constitution to acknowledge insanity of this magnitude. This type of denial is probably an unconscious protection mechanism that activates in order to prevent a nervous breakdown.

According to Dr. Goleman, this denial may have, at some point, assisted in our growth. He wrote, "In our evolution, our survival as a species may have hinged in part on our ability to select shrewdly, and to deceive ourselves just as shrewdly. But the capacity of the unconscious mind to pilot the conscious can backfire. ... We fall prey to blind spots, remaining ignorant of zones of information we might be better off knowing, even if that knowledge brings some pain."

Professor Marrs summed it up this way: "Truths that rattles one's nerves and unsettles one's inner security is truth that must be rejected. It simply must, or many believe they might go insane—or suffer an emotional breakdown. ... Eventually we must either reject it outright and refuse to listen to the truth, or finally, reluctantly, we are compelled to accept it, along with all the horrible complications and ramifications that come with the acceptance."

Apparently even some TIs don't believe the government is carrying this out. There are probably several reasons for this. First, they may be informants whose purpose is to mislead. They may just not be aware of the scope of it due to lack of study. Or, they may not want to admit that it's really that bad. In other words, *denial*.

"Even amongst TIs there is the perception in certain areas that our government would not do this," announced McKinney. She explains this denial as a case of "not recognizing reality." "First of all," she states, "if this were not being done by our government, congress would step in because of the ... thousands of complaints no doubt over the past 10-15 years, from citizens who *recognize* what's going on."

The masses of people in Germany and Russia were not aware that a dictatorship was being setup right in front of them. It can never happen here, right? It'd be all over the news... Entire countries have been under mind-control in past dictatorships. Professor Marrs contributes this to a type of denial that has occurred repeatedly throughout history. He asks, "If we are descending into a period of darkness in the world in which [a] totalitarian government controlled by the few elite is a reality ... then why is it hat that masses do not understand this? The answer," he says, "is that it is characteristic of a decaying civilization that those most closely associated with it are unconscious of the tragedy that has befallen them."

Some are fully aware that this program exists and how well funded, organized, and widespread it is. They're afraid that they'll become targeted if they attempt to expose it. Some who have tried to assist targeted people have become targets themselves. This is too big, widespread, powerful, and difficult to prove. So the popular and safe response is to pretend that the monster doesn't exist. Or pretend that it's not really a monster and join with it. The bottom line is that this is just too evil for most people to deal with.

"They don't want to lean the truth because it would be too discomforting—too alarming," wrote Professor Marrs. "If they were to accept the fact that much of what they think they know about their government and its leaders, the economy, and where society is heading are simply elaborate myths, concealing and insulating a hidden elite from opposition and exposure, they would then be faced with a horrible dilemma: What can they do about it? Even more scary is that it would also become mandatory that they ponder the question of what will happen to them if they try to do something about it."

If you're targeted, you may have trouble trying to convince someone who has spent most of their life watching sporting events, soap operas, corporate-controlled news, or playing video games, that this exists. The people who exist inside these ruts (most people) are under a form of mind-control. People, including some friends and family, may deny your claim that this exists. They may just not want to believe such a thing is possible. They may tell you, "you just need to snap out of it and get on with your life," or "you can't run away from your problems." Or similar remarks.

The truth is, *they* are the ones in denial. Or, they may be forced to lie about its existence as part of a gag order for a bogus investigation. It's been said that you can't wake someone up who pretends to be asleep. Friends or family may be forced into lying to your face, discrediting you by having you hospitalized, and participating in the harassment.

This is the *big evil* that everyone goes along with when they encounter it because they believe that once it passes, they'll be OK. Of course in the back of their minds they're thinking, "I hope it never comes for me or my family." My message to those who know what this is about is this: It is only getting worse. If you've read the previous chapters on history you know what they've done. Also, look into the facts related to 9/11. If you keep giving in, it will only be a matter of time before a family member becomes targeted. How will you feel when you are forced to harass your son, daughter or sibling?

Vested Interest

Some targeted people have been in touch with organizations that are devoted to assisting people who are being tortured and stalked. They either didn't respond, or they did not acknowledge that this program exists. I wrote a letter to the author of a book on stalking, who also happens to be the director of a large, popular anti-stalking organization. She responded by telling me that I should seek help from a psychiatrist. This women's reply also contained hints that I must have done something to provoke it and that I deserved it.

I checked out her website, and noticed that her organization is affiliated with a major Establishment institution. If they are affiliated with a corporation, they are probably receiving some type of funding (grants) from it. So, this would be an example of an organization that has a vested interest in covering up this epidemic. It is possible that her clinic is also being used for disinformation.

Summary

The program remains because the wealthy individuals who created it and use it, want it that way. All standard mainstream support organizations participate in the cover-up. This includes the legal and medical community, human rights organizations, local and federal law enforcement, and congress. In all likelihood, the national security act is preventing these organizations from intervening. Additionally, denial and fear may prevent some from recognizing the situation.

Chapter 28

Raising Awareness

There are a couple of considerations regarding what can be done about this program. There are things that a targeted individual can do to survive, and there things that can be done to help stop the program. In this chapter I will not include any suggestions on dealing with attacks.(*) Instead I'll focus on what I've concluded to be the best course of action to stop this program. This is also one of the shortest chapters because the message is simple: The reason it's called *Raising Awareness* is because the best thing that can be done at this point is to make people aware of it.

Fixing this problem and the underlying one which it's part of will require more than the suggestions given here. A carefully planned strategy (which this chapter doesn't provide) would be needed.(*1) So these are just some general guidelines. It could be argued that the infection has reached a stage where it can't be stopped. Nevertheless, stopping it should be strived for.

There is no quick way to stop this. It is just a symptom of a growing problem. The foundation for this dictatorship has been arranged—but they're not there, yet. In order to completely stop it, the movement toward a worldwide dictatorship must be stopped, and the stronghold that the financial elite have on America must be removed. However, it is at a stage in its development where it has gained such momentum that it will be almost impossible to stop by any means within the existing institutional framework.

Professor Marrs sums it up this way; "it may be too late for us to stop the men of the Secret Brotherhood," but it's "not too late to warn as many men and women as we can about the miserable fate that the hidden elite have in store for them."

There are still some decent people in government and the media, which is evident by media leaks and the government whistleblowers that appear on talk radio shows. If the people overseeing this were as influential as I've outlined, then it would seem there is very little that can be done to stop this. But this is not the case. This may sound like a contradiction, but eastern philosophy refers to this as "cleaving the opposites." It means that even though you understand that an evil exists and that it will continue, so too will your desire to stop it.

Earlier I wrote that because congress is under the control of the elite, a congressional investigation would be whitewashed. Although this is true, if enough people knew that this program existed and what it's part of—and acted—then maybe, maybe—that would provide the momentum to force congress to perform an *authentic* investigation. "The invisible government," commented Smoot, "is not, however, beyond the reach of the whole Congress," that is, "if the Congress has the spur and support of an informed public." He adds, "It is the people who must compel their elected representatives to make a thorough investigation of the Council on Foreign Relations and its interlock."

However, because the system is, literally, rotten to the core, before a legitimate investigation can be done, the primary mechanism that the global elite have used to control America—the Federal Reserve—must be eliminated. As previously demonstrated, this Invisible Government took over America beginning in 1913 when a private corporation was allowed gain control of the nation's money supply.

In the January 7, 1999 issue of *USA Today*, former US Secretary of Labor under Bill Clinton, Robert Reich stated, "The dirty little secret is that both houses of Congress are irrelevant. ... America's domestic policy is now being run by Alan Greenspan and the Federal Reserve, and America's foreign policy is now being run by the International Monetary Fund [IMF]."

The removal of the Federal Reserve will in turn eliminate the Tax-exempt Complex, which is funding their political movement. Smoot echoed this suggestion, writing, "One sure and final way to stop this great and growing evil," is to "eliminate the income-tax system which spawned it." He added, "It is the people who must compel Congress to deny administrative agencies of government the unconstitutional power of granting tax-exemption. It is the people who must compel Congress to submit a constitutional amendment calling for [a] repeal of the income tax amendment."

Alan B. Jones stated, "This change [to shut down the Federal Reserve] is extremely important, and will strike at the root of the elites' attack on our American society. Our free society is seen by them as the only entity on earth still capable of growing strong enough economically and socially to challenge their program for dominating the earth."

Fortunately, there is an effort underway to abolish the Federal Reserve, started by the late film producer Aaron Russo at www.freedomtofascism.com. I would consider his landmark documentary, *America: From Freedom to Fascism*, critical for anyone who wants to understand our situation, especially for targeted individuals who wonder how this program could exist.

If you are a TI, I suggest you sign up for Aaron's movement (or a similar one) to abolish the Federal Reserve. If you're part of the 9/11-truth movement, remember that whitewashes such as the 9/11 Commission are possible because congress is not in control. Therefore, 9/11 truthers should consider joining Aaron's movement. Once the Fed is removed, *then* congress can be persuaded to stop the Hidden Evil and related agendas.

In their book, *Access Denied: For Reasons of National Security*, Cathy O'Brien and Mark Phillips suggest repealing the 1947 National Security Act, which is abundantly used as an excuse to target the civilian population, as well as a pretext for a host of other crimes. I too believe this act should be revoked because it's obvious to me that it is being used as an excuse for the program, and to cover it up.

America must also be removed from the grip of the Council on Foreign Relations, the Trilateral Commission, the Bilderbergers, and any other supra-governmental agency that is promoting a worldwide government and a reduction of individual civil liberties. This means public exposure of these organizations. Smoot too recommends that we withdraw from all international, governmental, or quasi-governmental organizations, including the World Court, the United Nations, and UN specialized agencies.

Jones agreed, when he wrote, "The best-known instruments of control which they [the financial elite] are using are the United Nations, the International Monetary Fund, the World Bank, and ... the NAFTA ... GATT and ... World Trade Organization." He suggested that we withdraw from these organizations, and advised that the CFR should be recognized as a foreign entity operating within the United States.

Other critical issues which are intimately related include the events surrounding September 11th, and the Oklahoma City bombing. There must be a complete investigation into these events by individuals with no connections to the Establishment. Once again, I think the true power structure must be restored before this can happen. 9/11 was a major historical event which triggered drastic changes, and it is essential to find out what really happened.

Therefore, whether you're targeted or not, I would advise anyone who wants to learn what happened on 9/11 to visit 911truth.org, or 911sharethetruth.com. 9/11, the Oklahoma City bombing, the

Hidden Evil, and many other seemingly unconnected events are all part of a process.

Absent these things occurring, the best course of action is to raise awareness. Raising awareness will mean various things to each individual. Raising awareness is key because the one thing these people don't want is to be cast into public light. Anything this evil must operate in secrecy. "The evil hate the light," announced Dr. Peck.

"The one thing these conspirators cannot survive is exposure," stated Allen. He continued, "The Insiders are successful only because so few of their victims know what is being planned." Allen explained that they can operate only "in the dark," and "cannot stand the truthful light of day." Professor Marrs agreed, "Having a light shined on their repugnant activities causes these ... arrogant and despicable men to desperately scurry about seeking cover."

The mainstream media seems to be losing ground. People are turning to blogs and truth radio such as the *Genesis Communications Network* and the *Republic Broadcasting Network.* A massive change is happening where people instinctively know there is a major problem. The monopoly on the media does not have the same effect that it could have thanks to the internet. The internet is one of their biggest weaknesses. They are aware of this, and it would not surprise me if in the near future, they attempted to pass legislation which prevents people from posting information that exposes their activities. This legislation may be passed under the banner of "anti-hate" or "anti-terror" speech.

People who are not yet targeted (or don't yet realize they are), must be made aware that they are simply on the higher end of a sinking ship. Some of those who participate in this program, which is contributing to the enslavement of their children, may think that by participating they're "insiders", and as long as they go along, they'll be okay. This is false. They will be *used and disposed of*! Insiders are not people with a million dollars or so, but people of *tremendous* wealth. The only insiders are the financial elite.

"Unless you are an Insider, you are a victim," wrote Allen. He continued, "To the Insiders, the world is their country and their only loyalty is to themselves and their fellow conspirators. ... Being an American means no more to them than being an honorary citizen of Bali would mean to you. It has not bothered their consciences one iota that millions of your fellow human beings have been murdered..."

Each person needs to decide how to best use their personal gifts and resources. For some it may be handing out flyers or attending a rally. For others it means writing letters, posting information on message boards, blogging, putting up a website, or writing an article. But it also means educating yourself on this program and what it's part of, so that however you choose to raise awareness, you do so efficiently.

"The Job," described Allen, "is largely a matter of getting others to realize that they have been conned and are continuing to be conned. You must become the local arm of the word's largest floating university. But before you can go to work, pointing out these conspiratorial facts to others, you must know the facts yourself."

No doubt there are many informative documentaries on the state of affairs. I've only seen a small fraction of them since I began to research this agenda in early 2005. But out of the ones I've seen, some have been particularly fascinating. They include, *9/11: The Great Illusion*, by George Humphrey; *Martial Law 9/11: The Rise of the Police State*, by Alex Jones; and *America: From Freedom to Fascism*, by Aaron Russo.(*2)

Any organization which recognizes that a dictatorship is being setup is a potential ally. This includes 9/11 truth groups and truth radio talk shows.(*3) This also includes organizations which seek to restore the constitution and abolish the Federal Reserve. More than likely, some members of these organizations are targeted but are unaware of it.

Endnotes

* For a hotline and suggestions on how to cope, visit www.stopcovertwar.com.

*1 For more information on this, see the documentary, *The Money Masters: How the International Bankers Gained Control of America*, by William Still. Also see the book, *How the World Really Works*, by Alan B. Jones.

*2 These films can be purchased at www.infowars.com and www.freedomtofascism.com.

*3 www.wethepeoplefoundation.org, www.911truth.com, www.takebackwashington.com, www.republicbroadcasting.org, www.gcnlive.com.

Chapter 29

Commentary

Decent people are being tortured and murdered worldwide on a massive scale, by Non-lethal Weapons, and a colossal network of citizen informants who operate on behalf of their respective government. All key factions within the community, including everyone from seniors to children, engage in the ritualistic persecution of targeted people. Despite patriotic claims, informants are motivated by a selfish need for personal empowerment and adventure. Scapegoating also plays a role. The entire group functions as a single organism (Group Think) with a stunted mental capacity.

Installing and maintaining a dictatorship requires the creation of citizen informant networks, as illustrated by German and Russian dictatorships. Public participation is made possible by the use of a protocol that dictators have used as a pretext for mass persecution. Labeling someone a threat to national security, and associating them with undesirable groups have been standard procedures to destroy political enemies in past dictatorships.

Experts in the behavioral sciences and Non-lethal Weapons have worked with Think Tanks to promote this technology as part of a *New World Order*. Clearly, the program exists to destroy any opposition to their rule, or people they just don't like. The alleged creators of this program, congress, and the military, have openly declared that the weapons will be used on the civilian population. The program is thoroughly covered-up by mainstream organizations, most probably, for reasons of national security.

The Satanic traits that the program exhibits stem from high-level Satanic cults, such as The Order of Skull and Bones, and Bohemian Grove, which are fused with Think Tanks. Regardless of media whitewashes, these organizations work together to control nations and bring about a worldwide dictatorship (New World Order). A phased approach to stop the Hidden Evil is probably best; beginning first with the removal of the Federal Reserve. Joining organizations which are dedicated to this will prove helpful. In addition, joining organizations dedicated to truth which have already amassed a considerable amount of supporters, such as the 9/11-Truth movement, may be effective. Besides these things, the best thing to do is to raise awareness.

Chapter 30

Conclusion

Millions of plain-clothed citizens are now following and harassing people worldwide. The nature of the Gang Stalking and electronic attacks are consistent on a global scale. These groups have the complete support of local, state and federal governments. This includes everything from governmental organizations such as the IRS and federal law enforcement, down to small businesses in the local government.

After a person has been singled out, this network of informants is released on him like a pack of dogs. Large groups of people, consisting of everyone from senior citizens to children, stalk and harass targets in public. In both Russia and Germany, similar networks were fed lies by respected leaders operating on behalf of the state. Targets were labeled threats to national security, or associated with an undesirable group.

The present mass recruitment program appears to be USAonWatch. Presumably, before that it was a variety of state-sponsored informant networks, including the National Neighborhood Watch program. Groups and individual informants with no connection to any state program, also participate in public harassment, and are presumably responsible for the malicious acts committed against targets.

These citizen groups are supported by an intelligence division, which provides ammunition for public harassment by way of surveillance. Targeted people are also attacked with silent, through-the-wall Directed Energy Weapons. The motivation for these groups is most likely accomplished as it was in Germany and Russia—propaganda delivered by trusted authorities. Establishment-controlled organizations provide cover for the Hidden Evil.

This program of persecution is not carried out by the government you learned about in your history books. According to *Congressional Records*, that government was overthrown by a private corporation in 1913. Currently, supra-governmental Think Tanks such the CFR, TC and Bilderberg group, which are interlocked with Wall Street and the Tax-exempt Foundations, control America and other NATO nations. This interlock has been called *The Invisible Government* and *The Shadow Government*.

These Think Tanks are composed of multinational corporations, royalty, international banks, and people of tremendous wealth. Despite their humane appearance, they have declared war on citizens of the planet. They have funded dictatorships, instigated wars for profit, and have re-written history. They have also funded mind-control experiments and eugenics projects. They control mainstream media, which is their primary distribution center for lies and propaganda. Politics, industry, academia, and finance are also under their control. The president, whether he is a democrat or republican, is their puppet.

They create policy that is enacted worldwide, which is accomplished by circumventing congress and the voting public. The policy they set is filtered down into federal, state and local governments. The policy is equipped with propaganda used to promote it. It is enacted without public knowledge or approval. Influenced by convincing propaganda, there are people who carry out their policy with the best of intentions. The policy is given an official government "stamp" which gives it the appearance of having originated from the authentic government, which the general public still believes exists. The Hidden Evil is, in all likelihood, their policy.

The Hidden Evil is a local control mechanism, which is part of the installation of a worldwide socialist dictatorship. There is a Satanic/Psychopathic influence at the top of the control structure, which is part of the genesis and continuance of this program. Plain-clothed citizen informants are essential to establishing and maintaining a dictatorship. It is necessary for these citizen informant groups to be in place before an overt dictatorship is implemented.

Other indications of this impending dictatorship include: The merging of countries (European Union, North American Union, Pacific Union), the merging of currency (Euro and the impending Amero), the centralization of power under one governing body (UN promoted by the Think Tanks and Tax-exempt Complex), heavy surveillance of the population (already underway via anti-terror legislation), and a reduction in human rights (Patriot Act). More signs include a monopoly on media, industry and finance, which have already been accomplished.

The planet is in grave danger. The darkest period in history may be yet to come. The citizens who take part are pawns of a mechanism that is enslaving them. Most are unaware of the ramifications of their participation. The empowerment they receive seems to override critical thought. Their community leaders on the local level, such as police chiefs, mayors, councilmen, etc., have obviously endorsed this program. These people are deceived, corrupted, or intimidated. But like all large destructive processes, at some point, it will be known. This will not likely occur with mainstream publications, but with books such as this and related work.

About the Author

Mark was born in September of 1972 and grew up in Malden Massachusetts. He worked as a computer repairman for about ten years. Around late 2004 he began to realize that he was targeted. His interests include: dance (poppin), rustic furniture construction, gardening, current affairs, writing, and other activities. He presently serves as a farm-hand in Maine. His website is www.TheHiddenEvil.com.

Sources

Volume I Part I

-Overview

Above all the reader must—at least: *America's Secret Establishment: An Introduction to the Order of Skull and Bones, (Trine Day, 2003), Professor Antony C. Sutton*

-Introduction to the Financial Elite

Elite Clique Holds Power in the U.S: *The Shadows Of Power: The Council on Foreign Relations and the American Decline, (Western Islands, 1988), James Perloff*
Wars are started (and stopped): *Wall Street and the Rise of Hitler, ('76 Press, 1976), Professor Antony C. Sutton*
The real menace of our republic: *The Shadows of Power, James Perloff*
Congressman Lawrence P. McDonald wrote: *The Rockefeller File, (76 Press, 1976), Gary Allen*
The men at the top of the empire: *Dark Majesty: The Secret Brotherhood and the Magic of a Thousand Points of Light, (Rivercrest Publishing, 2004), Professor Texe Marrs*
From these they have typically proceeded to Harvard: *The Shadows of Power, James Perloff*
Five major groups: *Dark Majesty, Professor Texe Marrs*
significant in the determination of world affairs: *America's Secret Establishment, Professor Antony C. Sutton*
These include Think Tanks: *Ibid*
We should realize that many of these groups: *Dark Majesty, Professor Texe Marrs*
Perloff acknowledged this interlock: *The Shadows of Power, James Perloff*
This is a corporation in the nature of a foundation: *Foundations: Their Power and Influence, (Covenant House Books, 3rd printing, 1993), Rene Wormser*
[they] decide when wars should start: *Who's Who of the Elite, Robert Gaylon Ross Sr.*
Secret elitist groups always censor: *Trilaterals Over Washington, (August Corp, 1980), Professor Antony C. Sutton, Patrick M. Wood*
are not aware of its true objectives: *Ibid; America's Secret Establishment, Professor Antony C. Sutton; Rule by Secrecy, (Harper Collins, 2001), Jim Marrs*
These members are aware of only about 50%: *Who's Who of the Elite, Robert Gaylon Ross Sr.*
The existence of groups like the Trilateral Commission: *Rule by Secrecy, Jim Marrs*
If the top leadership of government and business: *Ibid*
Somewhere at the top of the pyramid: *The Invisible Government, (The Dan Smoot Report, Inc., 1962), Dan Smoot*
established some very special and highly influential: *None Dare Call It Conspiracy, (Concord Press, 1971), Gary Allen*
The Elite control our courts, the Pentagon: *Who's Who of the Elite, Robert Gaylon Ross Sr.*
not only the Central Intelligence Agency: *Rule by Secrecy, Jim Marrs*
The Rockefellers, Rothschilds: *Dark Majesty, Professor Texe Marrs*

-Trilateral Commission

The Trilateral Commission was founded: *Rule by Secrecy, Jim Marrs*
Its membership is composed of: *Who's Who of the Elite, Robert Gaylon Ross Sr.*
The Trilateral Commission was formally established: *The Shadows of Power, James Perloff*
With the blessing of the Bilderbergers: *Rule by Secrecy, Jim Marrs*
The organization is completely above ground: *America's Secret Establishment, Professor Antony C. Sutton*
Although its meetings are invitation-only: *Who's Who of the Elite, Robert Gaylon Ross Sr.*
It seeks to improve: *Trilaterals Over Washington, Professor Antony C. Sutton, Patrick M. Wood*
The paper suggested that leader: *Rule by Secrecy, Jim Marrs*
On 7 January 1977: *Trilaterals Over Washington, Professor Antony C. Sutton, Patrick M. Wood*
The January 16, 1977: *Ibid*
The same issue of The Washington Post: *Ibid*
The new President: *The Shadows of Power, James Perloff*
small handful of national leaders: *The Rockefeller File, Gary Allen*
In September 1974 Brzezinski: *Trilaterals Over Washington, Professor Antony C. Sutton, Patrick M. Wood*
a further vital and creative stage: *Ibid*
a national constitutional convention: *Ibid*
When Brzezinski refers to: *Ibid*
a skillful, coordinated effort to seize control: *With No Apologies: The personal and political memoirs of United States Senator Barry M. Goldwater, (Berkley, 1980), Senator Barry Goldwater*
The Crisis of Democracy: *Trilaterals Over Washington, Professor Antony C. Sutton, Patrick M. Wood*
corporate socialist takeover: *Ibid*
A new kind of fascism: *Ibid*
Like sheep going to slaughter: *Ibid*

-Council on Foreign Relations

If the Establishment is elusive: *The Shadows of Power, James Perloff*
Like its counterpart the Trilateral Commission: *Who's Who of the Elite, Robert Gaylon Ross Sr.*
The CFR and RIIA were originally: *The Shadows of Power, James Perloff*
The CFR puts out a publication: *Ibid*
They are located: *Who's Who of the Elite, Robert Gaylon Ross Sr.*
dominated by J.P. Morgan interests: *The Invisible Government, Dan Smoot*
Some other early/founding CFR members: *The Shadows Of Power, James Perloff; With No Apologies, Senator Barry Goldwater*
Funding for the CFR came: *Rule by Secrecy, Jim Marrs*
junior chapters: *The Shadows of Power, James Perloff*
dinner meetings: *The Invisible Government, Dan Smoot*
steering and advisory committees: *Who's who Of the Elite, Robert Gaylon Ross Sr.*
Let's start with the smoke, and mirrors: *Ibid*
superficially an innocent forum for academics: *Wall Street and the Rise of Hitler, Professor Antony C. Sutton*

The leadership of the invisible government: *The Invisible Government, Dan Smoot*
A number of individuals: *The Shadows of Power, James Perloff*
an inner core: *America's Secret Establishment, Professor Antony C. Sutton*
During its first fifty years: *The Rockefeller File, Gary Allen*
The power of the Council: *The Invisible Government, Dan Smoot*
Once the ruling members of the CFR: *The Shadows of Power, James Perloff*
submergence of US sovereignty: *The Rockefeller File, Gary Allen*
The Trilateral Commission doesn't secretly: *"W" Magazine, August 1978*
virtually an agency of the government: *Foundations: Their Power And Influence, Rene Wormser*
The C.F.R. is totally interlocked: *None Dare Call It Conspiracy, Gary Allen*
All of the organizations have: *The Invisible Government, Dan Smoot*
staffed almost every key position: *With No Apologies, Senator Barry Goldwater*
constitutions are irrelevant to the exercise of power: *The Shadows of Power, James Perloff*
School for Statesmen: *None Dare Call It Conspiracy, Gary Allen*
abolish the United States: *Ibid*
I am convinced that the Council: *The Invisible Government, Dan Smoot*
We shall have world government whether or (Warburg Quote): *The Rockefeller File, Gary Allen*
the creation of a world government: *The Shadows of Power, James Perloff*
not one American in five hundred: *Ibid*
one American in a thousand: *None Dare Call It Conspiracy, Gary Allen*

-The Bilderbergers
an international group: *Who's who Of the Elite, Robert Gaylon Ross Sr.*
leading political and financial figures: *None Dare Call It Conspiracy, Gary Allen*
The Bilderbergers are a group: *Rule by Secrecy, Jim Marrs*
The first official meeting: *None Dare Call It Conspiracy, Gary Allen*
The primary impetus: *Rule by Secrecy, Jim Marrs*
The Rockefellers and Rothschilds: *The Bilderberg Diary, (American Free Press, 2005), James P. Tucker*
The Bilderberg group is an organization: *Ibid*
in the form of the Bilderberger: *Dark Majesty, Professor Texe Marrs*
three interlocking Think Tanks: *Who's Who of the Elite, Robert Gaylon Ross Sr.*
Their meetings are financed: *None Dare Call It Conspiracy, Gary Allen*
The 'Bilderbergers' are another: *The Invisible Government, Dan Smoot*
close ties to Europe's nobility: *Rule by Secrecy, Jim Marrs*
As with the Trilateral Commission: *Ibid*
Those who adhere to the accidental: *None Dare Call It Conspiracy, Gary Allen*
sets global policy: *How the World Really Works, (ABJ Press, 1996), Alan B. Jones*
Bilderberg does not publish membership lists: *Who's Who of the Elite, Robert Gaylon Ross Sr.*
Each year the hosting government: *Ibid*
Decisions are reached: *None Dare Call It Conspiracy, Gary Allen*
they clear out all the guests: *Who's Who of the Elite, Robert Gaylon Ross Sr.*
A rather tight lid of secrecy: *None Dare Call It Conspiracy, Gary Allen*
When Prince Bernhard arrived: *Ibid*
Bilderberg keeps its news blackout: *The Bilderberg Diary, James P. Tucker*
Unlike their American counterparts: *Rule by Secrecy, Jim Marrs*

The press, naturally, is not allowed: *None Dare Call It Conspiracy, Gary Allen*
Reporters that are not invited: *The Bilderberg Diary, James P. Tucker*
Despite the fact that many highly: *Rule by Secrecy, Jim Marrs*
the news media are always present: *Who's Who of the Elite, Robert Gaylon Ross Sr.*
The Swiss Military to Protect International: *The Bilderberg Diary, James P. Tucker*
The 1994 meeting in Helsinki, Finland: *Ibid*
Versailles, France in 2003: *Ibid*
Gerald Ford attended Bilderberg: *Ibid*
In 1991, then Arkansas governor Bill Clinton: *Rule by Secrecy, Jim Marrs*
European super-state: *The Bilderberg Diary, James P. Tucker*
In some cases discussions do have an impact: *Rule by Secrecy, Jim Marrs*
a direct influence on the White House: *The Bilderberg Diary, James P. Tucker*
Most presidents have been members: *Ibid*
While the "new world order: *None Dare Call It Conspiracy, Gary Allen*
A single currency: *The Bilderberg Diary, James P. Tucker*
the word shadow government: *Ibid*
World Government, controlled by them: *Dark Majesty, Professor Texe Marrs*
may not be aware of the true intentions: *Rule by Secrecy, Jim Marrs*
Not everyone who attends: *None Dare Call It Conspiracy, Gary Allen*
neither Republican nor Democrat: *Dark Majesty, Professor Texe Marrs*
Gorbachev was influenced by the Bilderbergers: *The Bilderberg Diary, James P. Tucker*
He promises a new world order: *Ibid*
Hilary Clinton in: *Rule by Secrecy, Jim Marrs*

-The Federal Reserve System

From the earliest days: *None Dare Call It Conspiracy, Gary Allen*
Essential to controlling a government: *The Shadows of Power, James Perloff*
All those who have sought dictatorial: *None Dare Call It Conspiracy, Gary Allen*
an international banker: *With No Apologies, Senator Barry Goldwater*
a banker may receive political influence: *The Shadows of Power, James Perloff*
The bold effort the present bank: *Ibid*
In the early years of the Republic: *With No Apologies, Senator Barry Goldwater*
America heeded Jackson's warning: *The Shadows of Power, James Perloff*
A German banker named Paul Warburg: *Ibid*
The Panic of 1907: *Ibid*
an old hand at creating artificial panics: *None Dare Call It Conspiracy, Gary Allen*
Morgan the Great: *The Shadows of Power, James Perloff*
The Money Trust . . . caused the 1907 panic: *Ibid*
Heading National Monetary Commission: *Ibid*
Federal Reserve became law: *Ibid*
drafted on Jekyl Island: *None Dare Call It Conspiracy, Gary Allen*
allied with the Bilderberg group: *The Bilderberg Diary, James P. Tucker*
The Federal Reserve is privately owned: *The Shadows of Power, James Perloff*
Probably 90% of the US citizens: *Who's who of The Elite, Robert Gaylon Ross Sr.*
only Congress may issue money: *The Shadows of Power, James Perloff*
It operates outside of the control of Congress: *With No Apologies, Senator Barry Goldwater*
Since these positions control the entire economy: *None Dare Call It Conspiracy, Gary Allen*

the 16th amendment was passed: *The Shadows of Power, James Perloff*
Initially, it was nominal: *Ibid*
the Federal Reserve would stabilize the economy: *Ibid*
the Fed has usurped the government: *Ibid*
the invisible government: *Ibid*
If the key to controlling a nation: *Ibid*
plan for international currency: *Ibid*
In 1987 Senator Jesse Helms: *Ibid*
Many historians would have us believe: *Ibid*

-Tax-exempt Foundations
Foundations were originally created: *Foundations: Their Power and Influence, Rene Wormser*
By the time the income tax became law: *The Shadows of Power, James Perloff*
predominately tax avoidance: *The Invisible Government, Dan Smoot*
channeling of their fortunes into: *The Shadows of Power, James Perloff*
They have a power comparable to political patronage: *Foundations: Their Power and Influence, Rene Wormser*
a socialistic, one-world government: *Ibid*
Commission on Industrial Relations: *Ibid*
Supreme Court Justice Louis D. Brandeis: *Ibid*
Control is being extended: *Ibid*
the findings of 1915 are still significant: *Ibid*
foundations have permitted themselves to be infiltrated: *Ibid*
Congressman Cox died during the investigation: *The Invisible Government, Dan Smoot*
was a pathetic whitewash: *Ibid*
subversive in the extreme sense: *Foundations: Their Power And Influence, Rene Wormser*
"unfinished business" of the defunct Cox: *The Invisible Government, Dan Smoot*
CFR came under official scrutiny: *The Shadows of Power, James Perloff*
Other organizations which came under investigation: *How the World Really Works, Alan B. Jones*
A comfortable merger with the Soviet Union: *Ibid*
beyond the reach of a mere committee of the Congress: *The Invisible Government, Dan Smoot*
The net result of these combined efforts: *Foundations: Their Power And Influence, Rene Wormser*
an agency of the United States Government: *The Shadows of Power, James Perloff*
an overwhelming amount of evidence: *None Dare Call It Conspiracy, Gary Allen*
often act in concert with each other: *Foundations: Their Power and Influence, Rene Wormser*
Congressman Reece made a final report: *The Invisible Government, Dan Smoot*
Reece details the infiltration into the government: *Ibid*
Finally Reece concluded that: *Ibid*
When their activities spread: *Foundations: Their Power And Influence, Rene Wormser*
The power of the individual large: *Ibid*
Every significant movement to destroy: *The Invisible Government, Dan Smoot*

create for itself a favorable press: *Foundations: Their Power and Influence, Rene Wormser*
As I see it, the foundations: *The Invisible Government, Dan Smoot*
Perloff arrives at the same conclusion: *The Shadows of Power, James Perloff*
world government is necessary: *Foundations: Their Power and Influence, Rene Wormser*
devoted to a socialist one-world system: *The Invisible Government, Dan Smoot*
socialist and related political movements: *Foundations: Their Power And Influence, Rene Wormser*
moral relativity: *Ibid*
the brutal MKULTRA experiments: *The Search For The Manchurian Candidate: The CIA and Mind Control, (W. W. Norton & Company, 1991), Jonathan Marks*

-Commentary
On August 30, 2003 Congressman Ron Paul: *www.libertythink.com*
wish to set up their world dictatorship: *How the World Really Works, Alan B. Jones*

Volume I Part II

-Centralized Control of History, Media, and Academia
Control of Media
Rockefeller gang's plans for monopolistic: *The Rockefeller File, Gary Allen*
investigative new shows like Sixty Minutes: *The Shadows of Power, James Perloff*
The so-called founders of: *None Dare Call It Conspiracy, Gary Allen*
Congressional Record Volume 54: *The Shadows of Power, James Perloff*
The CFR needs to reach the mass audience: *The Invisible Government, Dan Smoot*
Public opinion is manufactured: *The Rockefeller File, Gary Allen*
CFR's influence in the mass media: *Ibid*
The Associated Press, New York Times: *Ibid*
The CFR could not accomplish: *Who's Who of the Elite, Robert Gaylon Ross Sr.*
CFR's ventriloquists: *The Rockefeller File, Gary Allen*
We are grateful to The Washington Post: *Chronological history of the new world order, Dennis L. Cuddy, PhD*
Other outlets allegedly under the control: *None Dare Call It Conspiracy, Gary Allen*
It is through the press and the media: *Dark Majesty, Professor Texe Marrs*
official landscape painters: *None Dare Call It Conspiracy, Gary Allen*
the engineering of consent: *The Rockefeller File, Gary Allen*
Establishment's managed news curtain: *None Dare Call It Conspiracy, Gary Allen*

Control of History and Education
According to some: *The Shadows of Power, James Perloff*
controlling or influencing the public press: *Foundations: Their Power And Influence, Rene Wormser*
Professor Kenneth Colegrove: *Ibid*
The impact of foundation money upon: *Ibid*
an enormous impact on education: *Ibid*
subservient to the wealth: *With No Apologies, Senator Barry Goldwater*
the historical blackout: *The Invisible Government, Dan Smoot*
a conscious distortion of history: *Foundations: Their Power And Influence, Rene Wormser*

Twentieth-century history: *Wall Street and the Rise of Hitler, Professor Antony C. Sutton*
the gullibility of the American: *America's Secret Establishment, Professor Antony C. Sutton*
The rewriting and deceitful misinterpretation: *Dark Majesty, Professor Texe Marrs*
so-called "accrediting" organizations: *Foundations: Their Power And Influence, Rene Wormser*
pet project of foundation managers: *Ibid*
Through foundations controlled: *Wall Street and the Rise of Hitler, Professor Antony C. Sutton*
Books critical of the official: *Trilaterals over Washington, Professor Antony C. Sutton, Patrick M. Wood*
historians that do not adhere: *The Shadows Of Power, James Perloff*
Where will you find a college administration: *The Invisible Government, Dan Smoot*
During the past one hundred years: *America's Secret Establishment, Professor Antony C. Sutton*
do not pour money into local school board races: *The Rockefeller File, Gary Allen*
rewrite the past: *The Shadows Of Power, James Perloff*
a conditioning mechanism: *America's Secret Establishment, Professor Antony C. Sutton*
Few of us are aware: *Dark Majesty, Professor Texe Marrs*

-Wall Street Funded Communists
Western textbooks on Soviet economic: *America's Secret Establishment, Professor Antony C. Sutton*
In the Bolshevik Revolution: *None Dare Call It Conspiracy, Gary Allen*
Jacob Schiff, the head of Kuhn: *The Shadows Of Power, James Perloff*
According to his grandson John: *None Dare Call It Conspiracy, Gary Allen*
emergence of a communist dictatorship: *Dark Majesty, Professor Texe Marrs*
Averell Harriman was a director: *Wall Street And The Rise Of Hitler, Professor Antony C. Sutton*
estimated even by Jacob's grandson: *The Shadows Of Power, James Perloff*
Congressman Louis McFadden: *None Dare Call It Conspiracy, Gary Allen*
Now our textbooks tell us that the Nazis: *America's Secret Establishment, Professor Antony C. Sutton*
But obviously these men: *The Rockefeller File, Gary Allen*
From the days of Spartacus: *Dark Majesty, Professor Texe Marrs*
On may 1st, 1918: *Wall Street and the Rise of Hitler, Professor Antony C. Sutton*
The Bolsheviks were not a: *None Dare Call It Conspiracy, Gary Allen*
Having created their colony in Russia: *The Rockefeller File, Gary Allen*
Probably no name symbolized: *The Shadows of Power, James Perloff*
Chase Manhattan built a truck factory: *With No Apologies, Senator, Barry Goldwater*
American technology helped: *The Shadows of Power, James Perloff*
Wall Street continued to aid the Russian: *None Dare Call It Conspiracy, Gary Allen*
these American capitalists: *Wall Street and the Rise of Hitler, Professor Antony C. Sutton*
Eaton Joins Rockefellers: *The Shadows of Power, James Perloff*
W. Averill Harriman was made: *Dark Majesty, Professor Texe Marrs*
Sutton quotes a report by Averell Harriman: *The Rockefeller File, Gary Allen*
based on assiduous research: *The Shadows of Power, James Perloff*

The synthesis sought by the Establishment: *America's Secret Establishment, Professor Antony C. Sutton*

-Pearl Harbor was Allowed to Happen

The Council on Foreign Relations played: *The Shadows of Power, James Perloff*
U.S. intelligence cracked the radio code: *Ibid*
these intercepts were routinely sent: *Ibid*
Captain Johan Ranneft: *Ibid*
a Japanese submarine was sunk: *The Times, US drew first blood at Pearl Harbor, August 31, 2002, Nicholas Wapshott*
The captain of the USS Ward: *The Shadows of Power, James Perloff*
Outerbridge sent a report: *The Times, US drew first blood at Pearl Harbor, August 31, 2002, Nicholas Wapshott*
Prior to the attack, Admiral Richardson: *The Shadows Of Power, James Perloff*
evidence to back US claims that it fired first: *BBC, Japanese Pearl Harbor sub found, August 29, 2002*
they found a Japanese midget submarine: *Chicago Sun-Times, Japanese sub sunk before attack found in Pearl Harbor, August 29, 2002*
no doubt the attack took place: *The Times, US drew first blood at Pearl Harbor, August 31, 2002, Nicholas Wapshott*
The Roberts Commission: *The Shadows of Power, James Perloff*
Supreme Court justice Owen Roberts: *Ibid*
At the court-martials: *Ibid*
repressed the results: *Ibid*
Despite testimony from Lieutenant Outerbridge: *The Times, US drew first blood at Pearl Harbor, August 31, 2002, Nicholas Wapshott*
Incriminating memoranda: *The Shadows of Power, James Perloff*
2,390 people dead, 1,178 wounded: *BBC, Japanese Pearl Harbor sub found, August 29, 2002*
The council on Foreign Relations has heavy: *The Invisible Government, Dan Smoot*
All this [funding] was done legally: *Rule By Secrecy, Jim Marrs*
the 9/11 Commission chaired, vice chaired:
infowars.com/print/Sept11/cfr_whitewash_commission.htm
If they really wanted to defeat the Nazis: *Wall Street and the Rise of Hitler, Professor Antony C. Sutton*

-Wall Street Funded Nazis

part of authentic history: *Wall Street and the Rise of Hitler, Professor Antony C. Sutton*
General Motors, Ford, General Electric, DuPont: *Ibid*
The deal bringing Hitler into the government: *Rule by Secrecy, Jim Marrs*
The financing for Adolph Hitler's rise to power: *None Dare Call It Conspiracy, Gary Allen*
In 1939, on the eve of blitzkrieg: *The Shadows Of Power, James Perloff*
Other U.S. companies which contributed heavily: *The New Hampshire Gazette, Bush-Nazi Link Confirmed, Documents in National Archives Prove, October 10, 2003, John Buchanan*
A similar article appeared: *The Guardian, How Bush's grandfather helped Hitler's rise to power, September 25, 2004, Ben Aris & Duncan Campbell*
Prescott Sheldon Bush: *Dark Majesty, Professor Texe Marrs*

Prince Bernhard of the Netherlands: *Wall Street and the Rise of Hitler, Professor Antony C. Sutton*
a founding member of the Bilderbergers: *Rule By Secrecy, Jim Marrs*
The alliance between Nazi Germany: *The Rockefeller File, Gary Allen*
responsible for bringing the Nazis: *Wall Street and the Rise of Hitler, Professor Antony C. Sutton*
German bankers on the Farben Aufsichsrat: *Ibid*
The two largest tank producers in Hitler's: *Ibid*
Newly discovered documents: *IBM Dealt Directly with Holocaust Organisers, The Guardian March 29, 2002*
alliance between IBM and the Reich: *The Village Voice, How IBM Helped Automate the Nazi Death Machine in Poland, March 27, 2002*
"accidental" or due to the "short-sightedness: *Wall Street and the Rise of Hitler, Professor Antony C. Sutton*
the United States accidentally played an important: *Ibid*
I.G. Farben directors had precise knowledge: *Ibid*
aided Naziism wherever possible: *Ibid*
All Honorable Men, published in 1950: *Ibid*
The evidence suggests there was a concerted effort: *Ibid*
Nazi industrialists were puzzled: *Ibid*
Their general attitude and expectation: *Ibid*
It was headed by the Council on Foreign Relations: *Ibid*
At the end of World War II: *Ibid*
So when we examine the Control Council: *Ibid*
None of the Americans were ever prosecuted: *The Shadows of Power, James Perloff*
were careful to censor any materials: *America's Secret Establishment, Professor Antony C. Sutton*
the very core of Naziism: *Wall Street and the Rise of Hitler, Professor Antony C. Sutton*

-The Great Depression was Deliberately Created
This new law [the Federal Reserve Act]: *The Shadows of Power, James Perloff*
That day of reckoning, of course, came in 1929: *Ibid*
Having built the Federal Reserve: *None Dare Call It Conspiracy, Gary Allen*
The Federal Reserve prompted: *The Shadows of Power, James Perloff*
despair [which] produced a willingness: *The Rockefeller File, Gary Allen*
It was not accidental: *The Shadows of Power, James Perloff*
Plummeting stock prices ruined: *Ibid*
History shows that the Wall Street: *How the World Really Works, Alan B. Jones*
For those who knew the score: *None Dare Call It Conspiracy, Gary Allen*
FDR is probably best remembered: *The Shadows of Power, James Perloff*
Longest and most severe economic: *Encyclopedia Britannica www.britannica.com*
To think that the scientifically: *None Dare Call It Conspiracy, Gary Allen*
Competition is a sin: *http://www.saidwhat.co.uk/quotes/favourite /john_d_rockefeller/competition_is_a_sin_9253*

-An Attempted Overthrow of the U.S. Government
offered up to 300 million dollars: *The Plot to Seize the White House, (Hawthorn Books, 1973), Jules Archer*
highest award for valor in: *http://www.cmohs.org*

Americans who plotted to seize the White House: *The Plot to Seize the White House, Jules Archer*
Heavy contributors: *Ibid*
Liberty League was the primary: *http://en.wikipedia.org/wiki/Business_Plot*
Most papers suppressed the whole story: *The Plot to Seize the White House, Jules Archer*
distortion, suppression, and omission: *Ibid*
struck a low blow at Butler: *Ibid*
string of denials, or ridicule: *Ibid*
It's a joke-a publicity stunt: *Ibid*
attempted to ridicule Butler: *Ibid*
Philadelphia Record and the New York Post: *Ibid*
He is a man of unquestioned sincerity: *Ibid*
the Senate committee expired: *http://en.wikipedia.org/wiki/Business_Plot*
no one involved in the plot had been prosecuted: *The Plot to Seize the White House, Jules Archer*
Portions of Butler's story were corroborated: *http://en.wikipedia.org/wiki/Business_Plot*
The Congressional Committee report confirmed Butler's testimony: *Ibid*
Roger Baldwin: *The Plot to Seize the White House, Jules Archer*
Time Magazine: *Ibid*
Stopped dead in its tracks when it got near the top: *Ibid*
too important politically: *Ibid*
There was no doubt that General Butler was telling the truth: *Ibid*
The founding president of the CFR: *The Shadows of Power, James Perloff*
Thomas W. Lamont was a founding member of the CFR: *Ibid*
one of the most fantastic plots in American history: *The Plot to Seize the White House, Jules Archer*
thwarted the conspiracy to end democratic government in America: *Ibid*
The Communists came to power: *None Dare Call It Conspiracy, Gary Allen*

Volume II Part I

-Introduction to the Hidden Evil

The program has been operational: *Microwave Harassment and Mind-Control Experimentation, Julianne McKinney*
Julianne McKinney is a former: *MHME, Julianne McKinney*
number one listing on Google: *Dr. Reinhard Munzert interviewed on The Investigative Journal, May 17, 2006, Topic-Directed Energy Weapons, www.Rbnlive.com*
particularly her interview: *Leuren Moret interviewed on Vancouver Co-op Radio CFRO 102.7 FM, August 8th 2005, Topic-The Technology of Political Control, www.coopradio.org*
"Mobbing" is the purposeful and strategic: *Uncovering the Truth about Depleted Uranium, Leuren Moret*
psychological terror: *Mobbing: Emotional Abuse In The American Workplace, (Civil Society Publishing, 2nd printing 2002), Noa Davenport, Ph.D., Ruth Distler Schwartz, and Gail Pursell Elliott*

-Tactics
The methods reportedly employed: *MHME, Julianne McKinney*
Zersetzung was developed to destroy: *Stasiland: True Stories from Behind the Berlin Wall, (Granta Books, 2003), Anna Funder*
In Germany this is called Zersetzung: *The Investigative Journal, Dr. Reinhard Munzert*
The German word Zersetzung: *Stasiland, Anna Funder*
frequency and duration: *Mobbing by Noa Davenport, Ph.D., Ruth Distler Schwartz, and Gail Pursell Elliott*

Basic Protocol
There is a basic protocol that they begin with: *Julianne McKinney interviewed on The Investigative Journal, April 19th, 2006, Topic-Directed Energy Weapons and Gang Stalking, www.Rbnlive.com*
double-folded strategy: *The Investigative Journal, Dr. Reinhard Munzert*
According to the DOJ: *Wired NewsWire, Big Business Becoming Big Brother, Kim Zetter, August 09, 2004*

Mental Health System
have used doctors to help abuse and discredit people: *Journey Into Madness: The True Story of Secret CIA Mind Control and Medical Abuse, (Bantam, 1990), Gordon Thomas*
psychiatrist and psychologists appear: *The Investigative Journal, Julianne McKinney*
enemies of the dictatorship: *Dark Majesty, Professor Texe Marrs*
The APA's refusal to acknowledge: *MHME, Julianne McKinney*
The first edition of the DSM: *American Psychiatric Association on Wikipedia.com; The Search for the Manchurian Candidate, Jonathan Marks; Mass Control: Engineering Human Consciousness, (Adventures Unlimited Press, 2003), Jim Keith*
brilliant cover up operation: *Microwave Mind Control: Modern Torture and Control Mechanisms Eliminating Human Rights and Privacy, September 25, 1999, Dr. Rauni Leena Kilde, MD*
first think of paranoia and schizophrenia: *The Investigative Journal, Dr. Reinhard Munzert*
It's called the Diagnostic Statistical Manual: *Annie Earle interviewed on The Investigative Journal, May 3, 2006, Topic-Experimentation, www.Rbnlive.com*
The patients that started coming: *Ibid*
has the ability to covertly assassinate U.S. citizens: *Nexus Magazine, How The NSA Harasses Thousands Of Law Abiding Americans Daily By The Usage Of Remote Neural Monitoring (RNM), April/May 1996, John St. Clair Akwei*

Surveillance
Similarly, German and Russian security: *Stasi: The Untold Story of the East German Secret Police, (Westview Press, 2000), John O. Koehler; The Persecutor, (Fleming H. Revell Company, 1973), Sergei Kourdakov*
bases of operation and training: *MHME, Julianne McKinney*
the people who move into these bases have been linked: *Ibid; Co-op Radio, Leuren Moret*
In order to target someone it requires: *The Investigative Journal, Julianne McKinney*
neighbors, friends and family are then co-opted: *Co-op Radio, Leuren Moret*
This surveillance is apparently done: *The Investigative Journal, Julianne McKinney*

The Life Assessment Detector System (LADS): *The Professional Paranoid: How to Fight Back When Investigated, Stalked, Harassed, or Targeted, by Any Agency, Organization, or Individual, (Feral House, 1999), H. Michael Sweeney*
Some of these flashlights display images in 3D format: *M2 Presswire, Handheld through-wall radar delivers unique 3D view that can revolutionize security, June 7, 2005; Design News, Radar device peeks through wall, August 6, 2001*

Character Assassination
It was planned destructiveness: *The Search for the Manchurian Candidate, Jonathan Marks*
the usual rumors spread: www.raven1.net; www.covertwar.com
A major effort is spent: *Co-op Radio, Leuren Moret*
Deception is the name of the game: *Microwave Mind Control, Dr. Kilde*
The agency was authorized to conduct secret smear campaigns: *http://edition.cnn.com/SPECIALS/cold.war/experience/spies/spy.files/intelligence/stasi.html*
The Russian KGB: *Center for Strategic and International Studies (CSIS) at Georgetown University, Understanding The Solzhenitsyn Affair: Dissent and its control in the USSR, Ray S. Cline*
The essence of defaming your target: *Gaslighting: How To Drive Your Enemies Crazy, (Loompanics Unlimited, 1994), Victor Santoro*
The Stasi would destroy the character: *The Mirror (London, England), The Stasi Files, When the Berlin Wall fell, the East German secret police panicked and shredded their secret papers. Now, a new machine is beginning to reassemble, May 28, 2007, David Edwards*
There have been multiple magazine and newspaper reports: *Rocky Mountain News, Sex offender falls victim to arson neighbors harassed him for months, says castle rock man, July 1, 2004, Gabrielle Crist; Wisconsin State Journal (Knight Ridder Newspapers), Living, isolated, under an unshakable label, registered sex offenders find it difficult to build a life after prison, August 8, 2005, Rick Armon; Associated Press Writer (AP Online), Pushing Sex Offenders May Increase Dangers, Neighbors and bosses force them from their homes and jobs, Michael Hill, June 20, 2005; Daily News (Los Angeles, CA) Associated Press, Convicted child molester loses job, January 16, 1997*
driven out of multiple states: *Associated Press Writer (AP Online), Sex Offender Chased from Four States, January 17, 2003, Julie Ann Stephens*
Perverted Justice would start campaigns: *Toronto Star, Vigilantes Versus Pedophiles, August 8, 2004*
calling friends, neighbors, children, places of employment: *Ibid*

Sensitivity Programs (NLP)
a very powerful tool: *Introducing NLP: Psychological Skills for Understanding and Influencing People, (Thorsons Publishers, 1993), Joseph O'Connor and John Seymour*
There is a basic protocol: *The Investigative Journal, Julianne McKinney*
They will stalk the target for a while: *Co-op Radio, Leuren Moret*
It is done by creating a peak emotional state: *Introducing NLP, Joseph O'Connor and John Seymour*
A stimulus which is linked to and triggers a physiological: *Ibid*
Emotional Transference: *Subliminal Mind Control, John J. Williams*

This is the realm of phobias: *Introducing NLP, Joseph O'Connor and John Seymour*
Targets are constantly: *The Investigative Journal, Julianne McKinney*
Organizations known to have studied NLP include military intelligence: *NLP Today E-zine February 2006, http://www.nfnlp.com; FBI Law Enforcement Bulletin, August 2001 Issue, Subtle Skills for Building Rapport, Using Neuro-Linguistic Programming in the Interview Room, Vincent A. Sandoval, M.A., and Susan H. Adams, M.A; Lobster Magazine, Non-Lethality, June 1993, Armen Victorian; SCS Matters, LLC http://www.scs-enterprises.com*
set of techniques to modify behavior patterns: *Ibid (Lobster Magazine)*
In 1983, the NLP training group: *The Warrior's Edge, (Avon Books, 1992), colonel John B. Alexander, Janet Morris, Major Richard Groller*
According to his Bio: *www.platinumstudios.com*
patterns of behavior can be installed: *The Warrior's Edge, colonel John B. Alexander, Janet Morris, Major Richard Groller*
There has been some concern in the mental health: *Introducing NLP, Joseph O'Connor and John Seymour; Gale Encyclopedia of Medicine, Neuro-linguistic programming, Leonard C. Bruno*
possible uses and misuses of this technology: *NLP: The New Technology of Achievement, (Harper Paperbacks, 1996), Steve Andreas and Charles Faulkner*
Targets around the world: www.raven1.net; www.covertwar.com
the process of copying an emotional state: *Introducing NLP, Joseph O'Connor and John Seymour*

Space Invasion (Crowding)
Space invasion includes blocking: *Personal Notes of Mark M. Rich, www.TheHiddenEvil.com; www.raven1.net; www.stopcovertwar.com*
Prolonged crowding can have an extremely: *Human Ecology, High stress and low income: The environment of poverty, December 1, 2002, Clare Ulrich; A Literature Review Prepared for the Ministry of Social Policy (Ebook), Definitions of crowding and the Effects of Crowding on Health, Alison Gray; American Demographics, Crowded House?-how different ethnic groups react to crowds, December 2000, Alison Stein Wellner; Human Ecology, Crowded Homes Are Stressful, Regardless of Culture, September 22, 2000, Susan S. Lang*
Body language is more accurate and reliable: *Messages: The Communication Skills Book, (New Harbinger Publications, 1995), Matthew McKay, Ph.D., Martha Davis, Ph.D., Patrick Fanning*
For most North Americans the intimate zone: *Ibid; Secrets of Sexual Body Language: Understanding non-verbal communication, (Amorata Press), Martin Lloyed-Elliot*
This space invasion happens: *Personal Notes of Mark M. Rich, www.TheHiddenEvil.com*

Noise Campaigns
noise campaigns: *Co-op Radio, Leuren Moret*
most people are sensitive to very loud noises: *Future War, (St. Martin's Press, 2000), Colonel John B. Alexander*
habitual door slamming: *Co-op Radio, Leuren Moret*
Door slamming: *MHME, Julianne McKinney*
Blaring horns, sirens, [and] garbage disposal[s]: *Ibid*
loudly pace as they mimic: *Co-op Radio, Leuren Moret*
A number of individuals report that occupants: *MHME, Julianne McKinney*

pace in their apartment: *Co-op Radio, Leuren Moret*
helicopters is a common tactic: *List of mind control symptoms, March, 2003, Cheryl Welsh, www.mindjustice.org; Personal Notes of Mark M. Rich, www.TheHiddenEvil.com*
noise will be synchronized: *Ibid (Rich)*
Chronic exposure to even low-level noise is considered a health hazard: *Archives of Environmental Health, Chronic effects of workplace noise on blood pressure and heart rate, July 1, 2002, Claire C. Caruso; Environmental Health Perspectives, How earplugs can help your heart: health effects of noise pollution, March 1, 2002, Bob Weinhold; Newsletter-People's Medical Society, Now Hear This (detrimental effects of noise), April 1, 1999*
noise can produce high blood pressure: *Nutrition Health Review, The increase in noise pollution: what are the health effects? (The Harmful Effects of Noise), September, 22, 1996, Arlene, L. Bronzaft*
Noise has been known to cause hearing loss: *Pediatrics (American Academy of Pediatrics Committee on Environmental Health), Noise: a hazard for the fetus and newborn, October 1, 1997; Pediatrics for Parents, Fetal hearing loss. (from intrauterine exposure to noise), February 1, 1993*
Prolonged stress in general: *Agence France Presse English, Stress triggers miscarriage, study says, November 10, 2004; Daily Post (Liverpool, England), Innocent nurse tells of year-long battle to clear name (News), March 15, 2005; War at Home: Covert Action Against U.S. Activists and What We Can Do About It, (South End Press, 1989), Brian Glick; The Mirror (London, England), Boss Stalked Me (News), November 15, 1995*

Synchronization
echoing and mirroring are very powerful: *Secrets of Sexual Body Language: Understanding non-verbal communication, Martin Lloyed-Elliot*
mirroring to influence people: *The Warrior's Edge, Alexander, Morris, Groller*

Harassment Skits (Street Theatre)
clear their throat or cough: *Personal Notes of Mark M. Rich, www.TheHiddenEvil.com*
Some words carry with them a particular weight: *NLP: The New Technology of Achievement, Steve Andreas and Charles Faulkner; Gaslighting, Victor Santoro*
the subject's ongoing experience: *Patterns of Hypnotic Techniques of Milton H. Erickson, M.D., (Metamorphous Press, 1997), John Grinder, Judith Delozier, Richard Bandler*
verbal anchors: *The Warrior's Edge, Alexander, Morris, Groller*
Metamessages are suggestions hidden in a statement: *Messages: The Communication Skills Book, Matthew McKay, Ph.D., Martha Davis, Ph.D., Patrick Fanning*
The tone, volume, and rhythm of specific words in the sentence are changed: *Ibid; Patterns of Hypnotic Techniques of Milton H. Erickson, M.D. Richard Bandler, John Grinder; Sunday Mercury (Birmingham, England), Fear, coercion and control-tactics used to recruit members, August 8, 1998*
It's hard to defend against the anger and disapproval: *Messages: The Communication Skills Book, Matthew McKay, Ph.D., Martha Davis, Ph.D., Patrick Fanning*
Imbedded commands serve the purpose of making suggestions: *Patterns of Hypnotic Techniques of Milton H. Erickson, M.D. Richard Bandler, John Grinder*
recurrent negative comments by strangers that: *MIIME, Julianne McKinney*

My friends tell me to feel comfortable: *Patterns of Hypnotic Techniques of Milton H. Erickson, M.D. Richard Bandler, John Grinder*
When we hear something, even from another conversation: *NLP: The New Technology of Achievement, Steve Andreas and Charles Faulkner*
foster your target's paranoia: *Gaslighting, Victor Santoro*
used on protesters at anti-corporate rallies: *San Francisco Bay Guardian, The new COINTELPRO, October 22, 2004*

Setups and Confrontations
Recurrent confrontations by unusually hostile strangers: *MHME, Julianne McKinney*
Seemingly homeless people: *Co-op Radio, Leuren Moret*
immediately complained to the building: *MHME, Julianne McKinney*
Reportedly, local police participate in stalking: *Ibid*
Cointelpro, there were routine setups: *War at Home, Brian Glick*
frame innocent people for crimes: *Sunday Herald, The FBI used this guy to frame men for murders they did not commit, December 7, 2003, Noel Young*
In case after notorious case: *The Franklin Cover-up: Child Abuse, Satanism, and Murder in Nebraska, (AWT Inc., 2005), former Senator John W. DeCamp*

Stores and Restaurants
Store and restaurant staff: *Personal Notes of Mark M. Rich, www.TheHiddenEvil.com*
A variation of these tactics: *Stasiland, Anna Funder*

Friends and Family
friends and family may be recruited: *Co-op Radio, Leuren Moret*
A number of individuals in touch with: *MHME, Julianne McKinney*
they will probably be forced to carry out: *Personal Notes of Mark M. Rich, www.TheHiddenEvil.com*

Thefts and Break-ins
When break-ins occur: *MHME, Julianne McKinney*
Santoro claims that these subtle: *Gaslighting, Victor Santoro*
Burglarizing your home is very, very common: *Co-op Radio, Leuren Moret*
Executive Order No. 12333: *War at Home, Brian Glick*
surreptitious break-ins, thefts, and sabotage: *Ibid*
Santoro advocates include filling: *Gaslighting, Victor Santoro*
always leave subliminal messages: *Techniques used to silence critics, Ted L. Gunderson FBI Senior Special Agent In Charge (Ret)*
if you are being subjected to a secret: *Subliminal Mind Control, (Consumertronics), John J. Williams*

Sabotage, Vandalism, and Staged Events
regularly experience acts of vandalism: *MHME, Julianne McKinney*
Ongoing computer problems: *Personal Notes of Mark M. Rich, www.TheHiddenEvil.com*
Some of this can be attributed to the E-bomb: *Talk of the Nation (NPR), Electromagnetic pulse bombs, March 14, 2003, Ira Flatow; The American Enterprise, Fear of a new weapon.(National Security), March 1, 2004, Eli Lehrer; AP*

Technology Writer (AP Worldstream), Pentagon could debut new weapons in Iraq, February, 18, 2003
pets die suddenly of mysterious illness: *MHME, Julianne McKinney*
Automobiles are one of the biggest targets: *Co-op Radio, Leuren Moret*
Vehicles invite peculiarly ferocious attacks: *MHME, Julianne McKinney*
staged accidents: *Ibid*
They also conduct staged accidents: *Co-op Radio, Leuren Moret*
"accidental" deaths: *MHME, Julianne McKinney*
Car brake-leads have been cut: *Stasiland, Anna Funder*

Traveling
surrounded by vehicles: *Personal Notes of Mark M. Rich, www.TheHiddenEvil.com; Road to Freedom shows, Eleanor White, www.shoestringradio.net*
City and state vehicles which may stalk: *Personal Notes of Mark M. Rich, www.TheHiddenEvil.com; List of Mind Control Symptoms, March, 2003, Cheryl Welsh, www.mindjustice.org*

Brighting
Brighting is an attack: *Personal Notes of Mark M. Rich, www.TheHiddenEvil.com*; www.raven1.net; *www.covertwar.com*
Among the most common non-lethal: *Winning the War, (St. Martin's Press, 2004), Colonel John B. Alexander*

Blacklisting
Targets who are employed: *MHME, Julianne McKinney*
Blacklisting was originally used as a foreign-policy tool: *Special to The Christian Science Monitor, Blacklisted by the bank, Sara B. Miller*
An entire industry has sprung up to produce: *The American Civil Liberties Union, The Surveillance-Industrial Complex: How the American Government Is Conscripting Businesses and Individuals in the Construction of a Surveillance Society, August 2004, Jay Stanley*
Federal law enforcement have sent these watch lists: *Wired NewsWire, Big Business Becoming Big Brother, August 9, 2004, Kim Zetter*
job applicants might have been denied work: *Ibid*
plain-clothed People's Brigade: *The Persecutor, Sergei Kourdakov*
To accomplish this Blacklisting: *Center for Strategic and International Studies (CSIS), Georgetown University, Understanding The Solzhenitsyn Affair: Dissent and its control in the USSR, Ray S. Cline*
their careers were destroyed: *Stasiland, Anna Funder*
to bring their finances down: *The Investigative Journal, Dr. Munzert*
financial state is destroyed: *Co-op Radio, Leuren Moret*
asked with accusing, or mocking tones: *Personal Notes of Mark M. Rich, www.TheHiddenEvil.com*
The harassment is masked: *Ibid*
Reporters and historians are also blacklisted: *The Shadows Of Power, James Perloff*
risk losing their jobs and being blacklisted: *Publishers Weekly editorial review at Amazon.com of book, Into the Buzzsaw: Leading Journalists Expose the Myth of a Free Press, by Kristina*

<u>Communications Interference</u>
Targets experience tampering: *MHME, Julianne McKinney*
mail is intercepted: *Co-op Radio, Leuren Moret*
During the former version of Cointelpro: *War At Home, Brian Glick*
inspected all mail in secret rooms: *Stasiland, Anna Funder*
disrupt the communication systems: *The Investigative Journal, Dr. Munzert*
targets usually receive lots of harassing telephone calls: *MHME, Julianne McKinney; Co-op Radio, Leuren Moret*

<u>Online</u>
Targets may also be stalked online: *Personal Notes of Mark M. Rich, www.TheHiddenEvil.com*
Carnivore, DCS1000: *Knight Ridder/Tribune News Service (Knight Ridder Newspapers), Modern snooping technology can dig deep, September 9, 2001, Dawn C. Chmielewski*
These cryptic attacks: *Subliminal Mind Control, John J. Williams*

<u>TI TV</u>
businesses conducting surveillance is increasing: *Business Wire, Statistics show rise in surveillance of workers, May 18, 2005*
people being Gang Stalked in the office: *The Investigative Journal, Julianne McKinney*
Covert workplace surveillance: *Denver Rocky Mountain News, Surveillance Helps Companies Catch Thieves Wireless Video Cameras, Transmitters Help Firms Keep An Eye on Staffers, April 30, 2001, Kyle Ringo*
Secret monitoring of employees: *Knight Ridder/Tribune News Service (Bridge News), The boss is watching as workplace surveillance grows, August 6, 1997, William S. Brown*
According to these publications: *Ibid; Denver Rocky Mountain News, Surveillance Helps Companies Catch Thieves Wireless Video Cameras, Transmitters Help Firms Keep An Eye On Staffers, April 30, 2001, Kyle Ringo; Who's watching the workplace? The electronic monitoring debate spreads to Capitol Hill (includes related article)(Special Issue) (Cover Story); Knight Ridder/Tribune Business News (The Orlando Sentinel), Privacy Experts Concerned by Employers' Increased Surveillance of Workers, July 23, 2003, Harry Wessel*

<u>Non-lethal Weapons</u>
briefcase-sized portable: *MHME, Julianne McKinney*
the law states that federal and local: *Chapter 170 of the Acts of 2004, An act relative to the possession of electronic weapons http://www.mass.gov/legis/laws/seslaw04/sl040170.htm*
These weapons or similar ones are currently being used: *Co-op Radio, Leuren Moret*
high-tech arms of the century: *Targeting the Human with Directed Energy Weapons, September 6, 2002, Dr. Reinhard Munzert*
this technology was outsourced to the FBI: *Co-op Radio, Leuren Moret*
anyone complaining about these systems were imagining: *The Investigative Journal, Julianne McKinney*
can cause nausea, fatigue, headaches, liquefy bowels: *U.S. News and World Report, EM Weapons (Wonder weapons), July 7, 1997, Douglas Pasternak; The Sunday*

Telegraph, Microwave gun to be used by US troops on Iraq rioters, September 19, 2004, Tony Freinberg, Sean Rayment
So-called acoustic or sonic weapons: *Ibid (Wonder weapons)*
Enemy soldiers might be confused: *Wired Digital, Surrender or We'll Slime You, February 1995, Mark Nollinger*
weapons are both land and space based: *Co-op Radio, Leuren Moret*
electrical grid throughout the country: *The Investigative Journal, Julianne McKinney*
Microwave DEWs will produce dizziness, burning, headaches: *Co-op Radio, Leuren Moret; The Investigative Journal, Julianne McKinney; The Investigative Journal, Dr. Munzert*
inflict pain in a highly focused fashion: *Ibid (McKinney)*
hit the heart, and the brain, and testes: *The Investigative Journal, Dr. Munzert*
cause deteriorating health, digestive problems, tremendous pain: *Co-op Radio, Leuren Moret*
Sleep disruption/deprivation: *MHME, Julianne McKinney*
We have information that this Secret Service: *The Investigative Journal, Dr. Munzert*
the Stasi File Authority began to investigate: *Stasiland, Anna Funder*
three of former East Germany's best-known dissidents: *BBC, Dissidents say Stasi gave them cancer, May 25, 1999*
his was tortured to death and given cancer: *The Investigative Journal, Dr. Munzert*
sustained high levels of radiation: *The People's State, Mary Fulbrook*
The Russian LIDA machine: *Eleanor White, www.raven1.net*
demonstrated a V2K success in the mid 70s: *The American Psychologist (Journal of the American Psychological Association), March 1975*
wireless" and "receiverless" communication of SPEECH: *Ibid*
The Walter Reed Army Institute of Research: *MHME, Julianne McKinney*
the inducement of auditory input: *Ibid*
type of torture: *The Investigative Journal, Dr. Munzert*
The intelligence agencies are capable of transmitting voices: *Virtual Government: CIA Mind Control Operations in America, Alex Constantine*
used on Iraqi soldiers: *Nexus, Military Use of Silent Sound (PSY-OPS WEAPONRY USED IN THE PERSIAN GULF WAR), Volume 5, Issue 6, October/November 1998, Judy Wall*
Subliminally, a much more powerful technology: *Ibid*
classified by the US Government: *Ibid*
Once you're engaged and you have a capability: *Cincinnati Post, U.S. Taps high-tech arsenal/pentagon likely to debut new weapons in Iraq, March 30, 2003*
the perfect crime: *The Investigative Journal, Dr. Munzert*

Support Groups
foster an informant's rise to respectability: *Department of History, Amherst College, The Stasi: New research on the East German Ministry for State Security, Catherine Epstein; The People's State,(Yale University Press, 2005), Mary Fulbrook*
Jacketing was often used during the old Cointelpro: *War at Home, Brian Glick*

-Informants
An Ancient Phenomenon
as far back as the Roman Empire:
http://ancienthistory.about.com/cs/rome/g/delatorinformer.htm; http://en.wikipedia.org/wiki/Delator

The fascist dictatorship of Portugal: *http://en.wikipedia.org/wiki/PIDE*
Other countries have used massive citizen informant networks: *BBC News, Czech collaborators to be named, March 6, 3003; BBC News, Czech communists guilty of harassment, February 12, 2002; http://www.espionageinfo.com, Czech Republic, Timothy G. Borden; The New York Times, Polish Assembly Votes to Release Files on Communist Collaborators, May 29, 1992; BBC News, New Polish archbishop in spy row, January 5, 2007; http://en.wikipedia.org/wiki/Sluzba_Bezpieczenstwa; http://en.wikipedia.org/wiki/StB; http://en.wikipedia.org/wiki/State_Protection_Authority; http://en.wikipedia.org/wiki/People%27s_Republic_of_Hungary; http://en.wikipedia.org/wiki/Ministry_of_Public_Security_of_Poland*
plain-clothed citizen informants were called IMs: *Stasiland, Anna Funder*
the result is nothing short of monstrous: *Stasi, John O. Koehler*
We have growing problems in our country: *The Persecutor, Sergei Kourdakov*

A Familiar Pattern

homeless people to white-collar workers: *Co-op Radio, Leuren Moret*
friends, co-workers and even family of targeted people: *Ibid*
are trained by this organization to harass: *Microwave Mind Control, Dr. Kilde*
Tens of thousands of persons in each [Metropolitan] area: *Nexus Magazine, How The NSA Harasses Thousands Of Law Abiding Americans Daily By The Usage Of Remote Neural Monitoring (RNM), April/May 1996*
Executive Order 12333 states: *War At Home, Brian Glick*
following and checking on subjects, Nexus Magazine: *Nexus Magazine, How The NSA Harasses Thousands Of Law Abiding Americans Daily By The Usage Of Remote Neural Monitoring (RNM), April/May 1996*
Even though TIPS was officially rejected by congress: *The American Civil Liberties Union, The Surveillance-Industrial Complex, Jay Stanley*
The Terrorism Information and Prevention System: *Sunday Morning Herald (SMH.com.au), US Planning to Recruit One in 24 Americans as Citizen Spies, July 15, 2002, Ritt Goldstein*
only under the most oppressive governments: *The American Civil Liberties Union, The Surveillance-Industrial Complex, Jay Stanley*
the current effort to build a colossal network: *The New American, TIPping off Big Brother, October 7, 2002, Steve Bonta*
1 million informants: *Washington Times, Planned Volunteer-Informant Corps Elicits '1984' Fears, July 15, 2002, Ellen Sorokin*
harness the power of the American people: *www.citizencorps.gov*
the IMs included doctors, lawyers, journalists: *Stasi, John O. Koehler*
Other questionable programs: *The Boston Globe, Building a Nation of Snoops, May 14, 2003, Carl Takei*
weed out: *U.S. Department of Justice, The Weed And Seed Strategy-Weed And Seed Topical Publication Series (Document NCJ 207498); U.S. Department of Justice, Weed And Seed Implementation Manual (Document NCJ 210242)*
Watching America with Pride: *The Boston Globe, Building a Nation of Snoops, May 14, 2003, by Carl Takei*
According to Catherine Epstein at the Department: *Department of History, Amherst College, The Stasi: New research on the East German Ministry for State Security, Catherine Epstein*

one in ten people will unofficially: *Techniques used to silence critics, Ted L. Gunderson FBI Senior Special Agent In Charge (Ret)*
In Germany the recruiters were given quotas: *Stasi, John O. Koehler*

They've Recruited the Youth
Children are now participating: *Personal Notes of Mark M. Rich, www.TheHiddenEvil.com*; *www.raven1.net*
families were recruited as IMs in East Germany: *The File, (Vintage Books, 1998), Timothy Garton Ash*
6% of the IMs were children: *Stasi, John O. Koehler*
These programs are probably offered in high school: *University Wire, Boston U. adopts Crime Watch program, September 6, 2001 Kerriann Murray; University Wire, Police enlist Ohio State U. students in war on crime, October 25, 2000, Ikenna D. Ofobike; The Atlanta Journal and Constitution, School police enlist aid of students in new crime watch, July 5, 2001, AHAN KIM; The Atlanta Journal and Constitution, Student-led crime watch on duty at high school, October 18, 2001, SHANDRA HILL; Sarasota Herald Tribune, Students keep the peace, October 16, 2000*
children were also recruited: *The People's State, Mary Fulbrook*
Youth Crime Watch: *www.ycwa.org; www.collegecrimewatch.org*

A Nation of Stalkers
USAonWatch is the face: *www.citizencorps.gov/programs/*
For more than 30 years: *www.bsheriff.net/documents/Usaonwatch.ppt.*
Citizen Corps is run by The Federal: *www.citizencorps.gov/about.shtm; www.in.gov/dhs/2488.htm*
The Neighborhood Watch is Homeland Security: *www.bsheriff.net/documents/Usaonwatch.ppt.*
literally encircled their everyday movements: *Man Without A Face, Markus Wolf, Anne McElvoy*
surrounded by unofficial informers: *The People's State, Mary Fulbrook*
people would be stalked by a network of IMs: *Stasiland, Anna Funder*
In 1974 the former Deputy Director: *Center for Strategic and International Studies (CSIS), Georgetown University, Understanding The Solzhenitsyn Affair: Dissent and its control in the USSR, Ray S. Cline*
help co-op citizens into participating in state-sponsored harassment: *Foreign Press Foundation, Ex-KGB and STASI Chiefs To Work Under Chertoff, December 16, 2004*
sponsored by the Department of Homeland Security: *VIPS-Volunteers In Police Service, www.policevolunteers.org*
in East Germany, the adults and children: *Stasiland, Anna Funder; Stasi, John O. Koehler*
In Russia, informants operated: *The Persecutor, Sergei Kourdakov; The Official KGB Handbook, (published by the USSR's Committee For State Security, translated and re-published by the Industrial Information Index)*
entire families who are citizen informants: *Chicago Sun-Times, Family of spies on run, With cover blown, pro informants `disappear', May 31, 1998, Pamela Warrick*

Secrecy
Under no condition was the public to know: *The Persecutor, Sergei Kourdakov*

When citizen informants in East Germany were recruited: *The People's State, Mary Fulbrook; The File, Timothy Ash*

Filtering
these mental blocking techniques: *Releasing The Bonds: Empowering People To Think For Themselves, (Aitan Publishing Company, 2000), Steven Hassan*
mindguard is an attention bodyguard, standing vigilant: *Vital Lies, Simple Truths, (Simon And Schuster, 1985), Dr. Daniel Goleman*
Hitler and his troop of amateur magicians: *Dark Majesty, Professor Texe Marrs*

Communication
IMs in East Germany communicated non-verbal cues: *Stasiland, Anna Funder*

Friends and Family
Friends and family of targeted people are sometimes: *Co-op Radio, Leuren Moret*
frequent maneuvers to generate tension: *War At Home, Brian Glick*
What we've studied and reported: *Co-op Radio, Leuren Moret*
put under siege: *Man Without A Face, Markus, Wolf, Anne McElvoy*
We know what you have been doing: *War At Home, Brian Glick*

-Structure
Full Government Complicity
ground crews that participate in patrols: *Co-op Radio, Leuren Moret*
Some of these informants: *Stasiland, Anna Funder*; *Man Without A Face, Markus Wolf, Anne McElvoy*
the internal army by which the government kept control: *Ibid (Funder)*
a criminal organization: *Stasi, John O. Koehler*
a key task of the Stasi: *The People's State, Mary Fulbrook*
To accomplish this: *Stasiland, Anna Funder*
Cointelpro has since been legalized: *War At Home, Brian Glick*
the FCC as a minimum: *The Investigative Journal, Julianne McKinney*
the FBI, CIA, and DOE are probably involved: *Ibid*
the military is ... very directly and heavily involved: *Co-op Radio, Leuren Moret*
This globally infiltrated organization: *Microwave Mind Control, Dr. Kilde*
The NSA gathers information on U.S. citizens: *Nexus Magazine, How The NSA Harasses Thousands Of Law Abiding Americans Daily By The Usage Of Remote Neural Monitoring (RNM), April/May 1996*

The Big Lie
There is an intelligence faction: *The Investigative Journal, Julianne McKinney*
each community has a Citizen Corps Council: *Citizen Corps, Guide for Local Officials, http://www.citizencorps.gov/cc/index.do*

What The Public Doesn't See
co-opting specific individuals in a targeted: *Co-op Radio, Leuren Moret*
attacks people with Directed Energy Weapons: *Ibid*
break-ins, vicious rumor campaigns, staged accidents: *MHME, Julianne McKinney; Co-op Radio, Leuren Moret*
very hierarchical: *Co-op Radio, Leuren Moret*
the use of the mafia and terrorists: *Microwave Mind Control, Dr. Kilde*

the FBI works with known criminals: *Sunday Herald, The FBI used this guy to frame men for murders they did not commit, December 7, 2003, Noel Young*
trained, and even protected terrorist: *The Times of India, CIA worked in tandem with Pakistan to create Taliban, March 7, 2001; Omaha World-Herald, Bin Laden once had U.S. support, experts say, September 17, 2001, Stephen Buttry and Jake Thompson; MSNBC, Bin Laden comes home to roost, His CIA ties are only the beginning of a woeful story, August 24, 1998, Michael Moran; WorldNetDaily, U.S. sent Afghanistan $125 million, September 15, 2001, Jon Dougherty; Newsweek, Alleged Hijackers May Have Trained At U.S. Bases, September 15, 2001, George Wehrfritz, Catharine Skipp and John Barry; MSNBC, The 'airlift of evil', November 29, 2001, Michael Moran; The Guardian, US helped Taliban to safety, magazine claims, January 21, 2002, Oliver Burkeman; The Washington Times, Report: bin Laden treated at US hospital, October 31, 2001, Elizabeth Bryant; The Guardian, FBI claims Bin Laden inquiry was frustrated, Officials told to 'back off' on Saudis before September 11, November 7 2001, Greg Palast and David Pallister*
The overt operations involve investigators: *Techniques used to silence critics, Ted L. Gunderson FBI Senior Special Agent In Charge (Ret)*

Hypothesis A
a theory ... which has to be supported by evidence: *America's Secret Establishment, Professor Antony C. Sutton*
Mr. Chris Morris: *www.m2tech.us*
Ms. Janet Morris: *Ibid*
the foreign and domestic use of Non-lethals: *MHME, Julianne McKinney*
The energy emitted from all of these [Microwave] weapons: *Ibid*
The Morris' have provided interviews and opinion pieces: *www.m2tech.us*
task force for the Council on Foreign Relations: *New York Times Magazine, The Quest For The Nonkiller App, July 25, 2004, Stephen Mihm*
Col. John B. Alexander, PhD:
http://original.platinumstudios.com/about/alexander_bio.php
the creation of this policy resulted from his non-lethals study for the: *Ibid*

-Purpose
New Weapons for a New World Order
Raytheon had developed the Active Denial System: *New York Times Magazine, The Quest For The Nonkiller App, July 25, 2004, Stephen Mihm*
weaponized versions of Valium and other drugs: *Ibid*
inked a contract ... with Platinum Studios: *USA Today (Gannett News Service), Top military advisor signs on for video games, July 3, 2003, By Mark Saltzman*
is recognized and acknowledged as the world's leading: *Market Wire, Colonel John B. Alexander, Ph.D. Secured as ShockRounds(TM) Advisor, March, 2004*
Maybe I can fix you: *The Washington Post Company, Mind Games, January 14, 2007, Sharon Weinberger*
able to protect national interests and values: *The Boston Globe, New Weapons For A New World Order, March 7, 1993*
Through many interlocking organizations: *The Invisible Government, Dan Smoot*
all extensively interlocked: *None Dare Call It Conspiracy, Gary Allen*
nonlethal technology with domestic law enforcement agencies: *Wired Digital, Surrender or We'll Slime You, February 1995 (Issue 3.02), Mark Nollinger*

vocal group of conspiracy theorists: *Harvard International Review, Optional Lethality-Evolving Attitudes Towards Nonlethal Weaponry, From The Future of War, Vol. 23 (2), Summer 2001*
Today the path to total dictatorship: *The Shadows Of Power, James Perloff*

<u>Hypothesis B</u>
[A] number of virtually identical complaints: *MHME, Julianne McKinney*
The awful truth is that the entire democratic: *Dark Majesty, Professor Texe Marrs*
If a revolution has indeed been accomplished: *Foundations: Their Power And Influence, Rene Wormser*
Here in the reality of socialism you have a tiny: *None Dare Call It Conspiracy, Gary Allen*
Article 4 of the Constitution of former East Germany: *Stasi, John O. Koehler*
Totalitarianism under any label spells: *Trilaterals Over Washington, Professor Antony C. Sutton, Patrick M. Wood*
In their pursuit of a new world order: *With No Apologies, former Senator Barry Goldwater*
If you have total government: *None Dare Call It Conspiracy, Gary Allen*
The key to modern history: *America's Secret Establishment, Professor Antony C. Sutton*
If one understands that socialism: *None Dare Call It Conspiracy, Gary Allen*
power-seeking billionaires in order to gain control over the world: *Ibid*
rather to seize control of the political state: *With No Apologies, former Senator Barry Goldwater*

<u>The Revolution is Over</u>
The most important thing of all is to remember: *Trilaterals Over Washington, Professor Antony C. Sutton, Patrick M. Wood*
socialist revolution by stealth: *Ibid*
All that can be hoped for is a counterrevolution: *Foundations: Their Power And Influence, Rene Wormser*
Revolution is always recorded as a spontaneous event: *America's Secret Establishment, Professor Antony C. Sutton*
IMs were to play an important role: *Department of History, Amherst College, The Stasi: New research on the East German Ministry for State Security, Catherine Epstein*
The CSIS report, which described: *Center for Strategic and International Studies (CSIS), Georgetown University, Understanding The Solzhenitsyn Affair: Dissent and its control in the USSR, Ray S. Cline*
The KGB's success depended on the extensive: *MHME, Julianne McKinney*
participatory dictatorship: *The People's State, Mary Fulbrook*
Every unfortunate society: *The New American, Informant nation, June 16, 2003*
capturing a percentage of the population: *The Investigative Journal, Julianne McKinney*
I'd like to go back to the underlying purpose for all of this: *Co-op Radio, Leuren Moret*
I think that once full control is established: *The Investigative Journal, Julianne McKinney*
Revolution is nearly accomplished: *Foundations: Their Power And Influence, Rene Wormser*

In order to solidify their power: *None Dare Call It Conspiracy, Gary Allen*
America to become part of a worldwide: *The Invisible Government, Dan Smoot*
If you want a national monopoly: *None Dare Call It Conspiracy, Gary Allen*
I would say that the persons who have realized: *The Investigative Journal, Julianne McKinney*
over Europe, and maybe all over the world: *The Investigative Journal, Dr. Reinhard Munzert*
These are global operations: *The Investigative Journal, Julianne McKinney*
carried out in all of the NATO countries: *Co-op Radio, Leuren Moret*
There is evidence that this is escalating: *The Investigative Journal, Annie Earle*
I've seen a tremendous expansion: *The Investigative Journal, Julianne McKinney*
They are ALREADY in every apartment block: *Microwave mind control, Dr. Kilde*
top management into a state-run program: *Stasi, John O. Koehler*
report directly to the police:
www.portlandonline.com/northportland/index.cfm?a=gcggb&c=dfjja
There was someone reporting to the Stasi on their fellows: *Stasiland, Anna Funder*
Without exception, one tenant in every apartment building: *Stasi, John O. Koehler*
Law enforcement, in partnership:
www.usaonwatch.org/Messages/AMessageFromTheDepartmentOfJustice.php
As groups continue to grow: *www.citizencorps.gov/programs/watch.shtm*
reinvigorated to increase the number of groups: *Citizen Corps, A Guide for Local Officials, www.citizencorps.gov/cc/index.do*
focal point for all Neighborhood Watch: *www.usaonwatch.org*
It's strange, before we were attacked: *The Investigative Journal, Dr. Reinhard Munzert*
pawns, shills, puppets: *None Dare Call It Conspiracy, Gary Allen*
dedication and effort to win that it took to destroy Hitler: *The Rockefeller File, Gary Allen*

-Motivational Factors

Intimidation and Blackmail

7% of the IMs were recruited by intimidation: *The People's State, Mary Fulbrook*

Voluntary Servitude

All human beings have these: *Personal Power II: The Driving Force, Anthony Robbins*
The surface reason is a means to an end: *Ibid*
it is not difficult to be biologically selfish: *Vital Lies, Simple Truths, Dr. Daniel Goleman*
But they believed that by participating they were special: *Stasiland, Anna Funder*
a member of a Russian Citizens Brigade: *The Persecutor, Sergei Kourdakov*
feeling important: *The People's State, Mary Fulbrook*
Whether it occurs among nations or among individuals: *Mobbing, Noa Davenport, Ph.D., Ruth Distler Schwartz, and Gail Pursell Elliott*
When these groups sprung up in Germany and Russia: *Stasiland, Anna Funder; Stasi, John O. Koehler; The Persecutor, Sergei Kourdakov*
A hallmark of the person as group member: *Vital Lies, Simple Truths, Dr. Daniel Goleman*
Individuals not only routinely regress: *People Of The Lie, (Simon and Shuster, Inc., 1983), Dr. M. Scott Peck*

Dr. Stanley Milgram quote: *www.Stanleymilgram.com*
Madness, said Nietzsche, is the exception: *Vital Lies, Simple Truths, Dr. Daniel Goleman*
Like individual evil, group evil is common: *People of The Lie, Dr. M. Scott Peck*
This is because a group is an organism: *Ibid*
mental homogeneity: *Vital Lies, Simple Truths, Dr. Daniel Goleman*
enemy creation: *People Of The Lie, Dr. M. Scott Peck*

Enemy Creation
It is almost common knowledge: *Ibid*
A group may implicitly demand of its members: *Vital Lies, Simple Truths, Dr. Daniel Goleman*
The first victim of groupthink is critical thought: *Ibid*
the behavior may seem disturbing when viewed by an outsider: *Betrayal Trauma: The Logic of Forgetting Childhood Abuse, (Harvard University Press, 1998), Dr. Jennifer J. Freyd*
a matter of laziness: *People Of The Lie, Dr. M. Scott Peck*
USAonWatch empowers: *www.citizencorps.gov/programs/watch.shtm*

Birds of a Feather
They often 'vulture,' lacking the courage: *Mobbing, Noa Davenport, Ph.D., Ruth Distler Schwartz, and Gail Pursell Elliott*
A number of individuals having surrendered their selves to something bigger: *Ibid*
Covert wannabees: *The Investigative Journal, Julianne McKinney*
You would think it would not be so easy to co-opt them: *Co-op Radio, Leuren Moret*
Specialized Groups: *People Of The Lie, Dr. M. Scott Peck*

Evil in Disguise
They live down the street-on any street: *Ibid*
How can they be evil and not designated: *Ibid*
Utterly dedicated to preserving their self-image of perfection: *Ibid*
Are crucial to understanding the morality of the evil: *Ibid*
Evil people are often destructive because they are attempting: *Ibid*

Ordinary Evil
Even civilians will commit evil with remarkable: *Ibid*
The point of the experiment is to see: *Harper's Magazine, The Perils of Obedience, December 1973, Stanley Milgram, http://home.swbell.net/revscat/perilsOfObedience.html*
"These predictions," stated Dr. Milgram "were unequivocally wrong.": *Ibid*
65% of the adult test subjects administered the full 450 volts: *Psychology Today Magazine (Document ID: 1989), The Man Who Shocked the World, March/April 2002, Thomas Blaas*
The level of obedience increased to has high as 85%: *Harper's Magazine, The Perils of Obedience, December 1973, Stanley Milgram*
ordinary people, simply doing their jobs, and without any particular hostility: *Ibid*
the satisfaction of pleasing the authority figure: *Ibid*
the experimenter did not threaten the subjects: *Ibid*
the responsibility belonged to the man who actually pulled the switch: *Ibid*

Obviously, you would have dissented: *Yale Alumni Magazine, When Good People Do Evil, January/February 2007, Philip Zimbardo '59PhD*
soar to over 90 percent: *Ibid*
authoritarian political movement: *Ibid*
some disturbing truths about human nature: *Psychology Today Magazine, The Man Who Shocked The World, March/April 2002, (Document ID: 1989), Thomas Blaas*
demonstrated with jarring clarity that ordinary individuals could be induced: *Ibid*
Milgram crafted his research paradigm to find out what strategies: *Yale Alumni Magazine, When Good People Do Evil, January/February 2007, Philip Zimbardo '59PhD*
If a system of death camps were set up in the United States: *The American Scientist, Milgram's Progress, July/August 2004, Robert V. Levine, http://www.americanscientist.org*
Primetime wanted to know if ordinary people: *ABC News Internet, Basic Instincts: The Science of Evil, January 3, 2007, Caroline Borge (abcnews.com)*
group will remain inevitably potentially conscienceless and evil: *People Of The Lie, Dr. M. Scott Peck*
Stanley Milgram quote*: www.Stanleymilgram.com*

- Potential Targets
Air Force official says non-lethal weapons: *Associated Press, Official Touts Nonlethal Weapons for Use, September 12, 2006, Lolita C. Baldor*
We are developing devices: *Peace Magazine, Zapping The Movement, Jun-Jul 1987, David Kattenburg*
harassed in multiple states and countries: *Sun Journal, New Bern, NC, Woman fears government zapping, September 28, 1992, Janet Blackman*
harassment campaigns are intergenerational: *MHME, Julianne McKinney*
No TI should look for a reason why this is going on: *The Investigative Journal, Julianne McKinney*
are not and never have been a threat: *Ibid*
absolutely terrible secret policy: *Co-op Radio, Leuren Moret*
reasons of national security blanket: *MHME, Julianne McKinney*
Similarly, in East Germany and Russia, people were labeled: *The Persecutor, Sergei Kourdakov; Stasi, John Koehler; Stasiland, Anna Funder*
other groups may emerge: *Future War, Colonel John B. Alexander*
The government's legitimate concern with national security: *The Franklin Cover-up, former Senator John Decamp*
There are no laws preventing this: *The Investigative Journal, Julianne McKinney*

Whistleblowers
Government whistleblowers or whistleblowers: *MHME, Julianne McKinney*
Leuren Moret: *Uncovering the Truth about Depleted Uranium, Leuren Moret*
The whistle-blower is forced out: *Virtual Government, Alex Constantine*

Protestors
The records show that the vast majority of the targets: *War At Home, Brian Glick*
40 people experienced headaches: *The Guardian, Peace women fear electronic zapping at base, Monday March, 10 1986, Gareth Parry*
trying to kill her with radio frequency weapons: *Associated Press, Government Wants Me Dead, April 15, 1995, (www.mindjustice.org)*

potential "terrorists.": *The Village Voice, J. Edgar Hoover Back at the 'New' FBI, Classified FBI Bulletin Reveals Tactics at Protests, December 4th, 2003, Nat Hentoff*
targeted due to uncovering some information:
www.arcticbeacon.citymaker.com/articles/article/1518131/38381.htm
demonstrators disrupting World Trade Organization: *Seattle Times, Crowd-control cookery: Microwaves among new non-lethal weapons, April 1, 2000, John Yaukey*
"neuter" them: *The Washington Post Company, Mind Games, January 14, 2007, Sharon Weinberger*
Parents and children of targeted individuals are also being: *MHME, Julianne McKinney*
killed one of her daughters to spare her: *Cleveland Plain Dealer, Psychiatrist Testifies At Mom's Hearing, June 28, 1991*
if she valued the lives of her children: *MHME, Julianne McKinney*
the Stasi would target families: *The People's State, Mary Fulbrook*

<u>Experimentation</u>
testing the latest in surveillance and Directed Energy Weapons: *The Investigative Journal, Julianne McKinney*
MKULTRA, a known mind-control project: *The Search for The Manchurian Candidate, Jonathan Marks*
about 23,000 people were traumatized: *Mass Control-Engineering Human Consciousness, Jim Keith*
Stasi burnt out many paper-shredding machines: *Stasiland, Anna Funder*
The paper trail for these experiments ended around 1984: *Ibid*
It would appear that the CIA's and FBI's Operations MKULTRA: *MHME, Julianne McKinney*

<u>Recruitment</u>
Both the East German Stasi and the CIA: *Stasiland, Anna Funder; The Search For The Manchurian Candidate, Jonathan Marks*

<u>Political Movement</u>
it has moved beyond experimentation: *The Investigative Journal, Julianne McKinney; Co-op Radio, Leuren Moret*
ANYONE can become a target: *Microwave mind control, Dr. Kilde*
many of them run surveillance operations for government agencies: *The American Civil Liberties Union, The Surveillance-Industrial Complex, August 2004, Jay Stanley*
many instances of individuals in the corporate environment being singled: *The Investigative Journal, Julianne McKinney*
there does seem to be a certain profile: *The Investigative Journal, Annie Earle*
There's a heavy predominance of those types: *The Investigative Journal, Julianne McKinney*
Others that have been targeted include: *Ibid (McKinney); Eleanor White, www.raven1.net; Suanne Campbell, www.stopcovertwar.com; Road To Freedom shows, Eleanor White www.shoestringradio.net; Co-op Radio, Leuren Moret; The Investigative Journal, Dr. Reinhard Munzert*
Mobbing targets are selected for these exact reason: *Mobbing, Noa Davenport, Ph.D., Ruth Distler Schwartz, and Gail Pursell Elliott*

Eugenics
there are considerably more women targeted than men: *The Investigative Journal, Julianne McKinney*
Families such as DuPont, Harriman, and Rockefeller: *Rule By Secrecy, Jim Marrs*
Commission on Population Growth: *The Rockefeller File, Gary Allen*
The Harrimans ... along with the Rockefellers funded: *Rule By Secrecy, Jim Marrs*
eugenics programs in Nazi Germany were organized and funded: *Mass Control-Engineering Human Consciousness, Jim Keith*
Eugenics work, under more politically correct names: *Rule By Secrecy, Jim Marrs*
Rockefellers and Carnegies constructed the Eugenics Records: *Mass Control-Engineering Human Consciousness, Jim Keith*

-Effects
force the individual to commit an act of violence: *MHME, Julianne McKinney*
The purpose is to completely isolate: *Co-op Radio, Leuren Moret*
Isolation of the individual from members: *MHME, Julianne McKinney*
induce a sense of perverted "loyalty" toward: *Ibid*
Your protocol follows you: *The Investigative Journal, Julianne McKinney*
All of these harassment techniques are recurrent: *Co-op Radio, Leuren Moret*
Suicides might also qualify as 'staged accidents,': *MHME, Julianne McKinney*

Symptoms
Because the group harassment for Mobbing and The Hidden Evil: *Mobbing, Noa Davenport, Ph.D., Ruth Distler Schwartz, and Gail Pursell Elliott*
other reported symptoms and situations are: *Co-op Radio, Leuren Moret; Microwave mind control, Dr. Kilde; The Investigative Journal, Dr. Reinhard Munzert; MHME, Julianne McKinney; The Investigative Journal, Julianne McKinney*
Many, many, many thousands, no doubt, are involved: *The Investigative Journal, Julianne McKinney*
are experiencing extreme traumatic stress: *The Investigative Journal, Annie Earle*
the victims experience extreme, unbelievable things: *Targeting the Human, Dr. Reinhard Munzert*
were probably more damaging, ultimately: *Man Without A Face, Markus Wolf*
For many, the discovery: *Bulletin of the Atomic Scientists, Stasi: The Untold Story of the East German Secret Police, July 1, 1999, Bourgholtzer, Frank*
Rapidly deteriorating health, generally of a digestive nature: *MHME, Julianne McKinney*
hot spot on the skull: *The Investigative Journal, Julianne McKinney*
Usually the victims experience pain: *The Investigative Journal, Dr. Reinhard Munzert*
People with mental disorders are stereotypically: *Center for the Advancement of Health, Stigma of mental illness still exists, Bernice A. Pescosolido, PhD, http://www.cfah.org*
Frankly I strongly recommend that you keep your: *The Investigative Journal, Julianne McKinney*
lunatic asylum: *The Investigative Journal, Dr. Reinhard Munzert*
harassment even continues inside hospitals: *MHME, Julianne McKinney; Cleveland Plain Dealer, Psychiatrist Testifies at Mom's Hearing, June 28, 1991*
If you have been diagnosed with a mental disorder: *Connecticut Clearinghouse-A program of wheeler clinic, Stigma and Mental Illness, (www.samhsa.gov)*

You may also be denied certain licenses: *PSYCHIATRIC STIGMA follows you everywhere you go-for the rest of your life (A warning from Lawrence Stevens, J.D), Lawrence Stevens, J.D.*
EAPs work for the company: *Mobbing, Noa Davenport, Ph.D., Ruth Distler Schwartz, and Gail Pursell Elliott*
it is better to be an ex-convict in the job market than someone: *PSYCHIATRIC STIGMA follows you everywhere you go-for the rest of your life (A warning from Lawrence Stevens, J.D), Lawrence Stevens, J.D.*

Results of the Hidden Evil
There are persons I have seen being targeted: *The Investigative Journal, Julianne McKinney*
6 million people came under scrutiny: *http://en.wikipedia.org/wiki/Stasi; Stasi, John O. Koehler*
have been known to go postal and kill themselves: *Mobbing, Noa Davenport, Ph.D., Ruth Distler Schwartz, and Gail Pursell Elliott*
a wave of postal killings, giving rise to the term: *Atlanta's Office Massacre, Rachael Bell, http://www.crimelibrary.com*
members of the intelligence community: *Virtual Government, Alex Constantine*
A jury awarded $5.5 million to the family of a woman: *Employment Practices Solutions, Inc (Articles/News Briefs), $5.5 Million Awarded in Sexual Harassment Case, November 6, 1996*
Mr Lee left a suicide note addressed to his mother: *The Guardian, Suicide of black worker 'caused by bullying,' January 9, 2001, Helen Carter*
unable to ever return to the workforce: *Mobbing, Noa Davenport, Ph.D., Ruth Distler Schwartz, and Gail Pursell Elliott*
the post office may have been responsible: *The Milwaukee Journal Sentinel, The Report links violence, local postal management, December 22, 1998, Georgia Pabst*
In 1999 a postal inspector named Calvin Comfort: *The Milwaukee Journal Sentinel, Postal inspector kills self near work, April 2, 1999, Jessica Mcbride*
Milwaukee postal inspector who investigated the fatal 1997: *Ibid*
A former postal worker who had been put on medical leave: *USA Today, Ex-postal worker commits suicide after 5 die in shooting, January 31, 2006*
Psychological harassment and intimidation: *Virtual Government, Alex Constantine*
Some may feel revengeful: *Mobbing, Ph.D., Ruth Distler Schwartz, and Gail Pursell Elliott*
refusal of the mass media to address this topic: *MHME, Julianne McKinney*
The U.S. postal service has conducted its own $4 million dollar study: *Boston Herald, Workplace violence called an epidemic, December 27, 2000, Tom Mashberg; The Atlanta Journal-Constitution, Study Calls 'Go Postal' Stereotype Mere Myth, September 01, 2000, Bob Dart*

Volume II Part II

-The Psychopathic Influence

Description of a Psychopath
Psychopaths, also called sociopaths: *Mask of Sanity, (C.V. Mosby Co.), Dr. Hervey Milton Cleckley; Without Conscience, (Simon and Schuster Inc, 1993), Dr. Robert D. Hare*

emotional impairment: *The Psychopath: Emotion and the Brain, (Blackwell Publishing, 2005), James Blair, Karina Blair, Derek Mitchell*
psychopaths don't feel emotion in a normal: *The Psychopathic Mind: Origins, Dynamics, and Treatment, (Jason Aronson Inc., 1988), Dr. J. Reid Meloy*
Generally, those who believe: *The Psychopath: Emotion and the Brain, James Blair, Karina Blair, Derek Mitchell*
These disorders are detected: *Ibid*
4% of the population: *Without Conscience, Dr. Robert D. Hare*
can be found in every race: *Ibid*
the term psychopath has been used: *Ibid*
no difficulty at all in looking anyone: *Mask of Sanity, Dr. Hervey Milton Cleckley*
Lying, deceiving, and manipulation: *Without Conscience, Dr. Robert D. Hare*
ability to rationalize their behavior: *Mask of Sanity, Dr. Hervey Milton Cleckley*
show a stunning lack of concern: *Without Conscience, Dr. Robert D. Hare*
counseling would be wasted: *Ibid*
reflect the wishes of the therapist: *The Psychopathic Mind, Dr. J. Reid Meloy*
Among themselves therapists: *People of the Lie, Dr. M. Scott Peck*
traditional therapeutic approaches: *http://en.wikipedia.org/wiki/Psychopathy*
considered sane: *Without Conscience, Dr. Robert D. Hare*
Dr. Scott Peck concludes: *People of the Lie, Dr. M. Scott Peck*

Where Are They?
The true difference between them: *Mask of Sanity, Dr. Hervey Milton Cleckley*
never go to prison or any other facility: *Without Conscience, Dr. Robert D. Hare*
Corrupt and callous politicians: *The Psychopath's Brain: Tormented Souls, Diseased Brains, Renato M.E. Sabbatini, PhD*
The experience of pleasure: *The Psychopathic Mind, Dr. J. Reid Meloy*
They love to have power: *Without Conscience, Dr. Robert D. Hare*
They are well aware that their mental makeup: *Political Ponerology: A science on the nature of evil adjusted for political purposes, (Red Pill Press, 2006), Andrew M. Lobaczewski*
learn how to detect others: *Ibid*
contacted by Nicole Kidman: *Psychopaths Among Us, Robert Hercz, www.hare.org/links/saturday.html*
any emotions which the primary psychopath: *http://en.wikipedia.org/wiki/Psychopathy*
using their charm: *Without Conscience, Dr. Robert D. Hare*
the typical psychopath will seem particularly: *Mask of Sanity, Dr. Hervey Milton Cleckley*
often witty and articulate: *Without Conscience, Dr. Robert D. Hare*

A Different Species
Some researchers agree: *Political Ponerology, Andrew M. Lobaczewski; Psychology Today Magazine, This Charming Psychopath, Robert Hare*
imagining the world as the psychopath: *Without Conscience, Dr. Robert D. Hare*
the world of pathological egotism and terror: *Political Ponerology, Andrew M. Lobaczewski*
The typical person: *None Dare Call It Conspiracy, Gary Allen*
intraspecies predators: *The Psychopath's Brain, Renato M.E. Sabbatini, PhD*

There is a class of individuals: *Psychology Today Magazine, This Charming Psychopath, Robert Hare*
what is missing in this picture: *Without Conscience, Dr. Robert D. Hare*
morally depraved individuals: *Antisocial Personality, Sociopathy, And Psychopathy, North Carolina Wesleyan College, http://faculty.ncwc.edu/toconnor/428/428lect16.htm*
create chaos for no reason: *Mask of Sanity, Dr. Hervey Milton Cleckley*
One day a scorpion and a frog: *http://xtiandefence.blogspot.com/2006/05/scorpion-and-frog-may-15.html*

Psychopaths in Politics
they are likely to exert themselves more than most: *People Of The Lie, Dr. M. Scott Peck*
[They] come from the very highest social strata: *None Dare Call It Conspiracy, Gary Allen*
We then usually reach the conclusion: *Political Ponerology, Andrew M. Lobaczewski*
Because their willfulness: *People Of The Lie, Dr. M. Scott Peck*
under stressing social situations: *The Psychopath's Brain, Renato M.E. Sabbatini, PhD*
Yet, in public, they impress us as men: *Dark Majesty, Professor Texe Marrs*
When psychopaths rule a society: *Political Ponerology, Andrew M. Lobaczewski*
When the controlling faction: *Ibid*
Congressman Lawrence P. McDonald: *The Rockefeller File, Gary Allen*
EVIL MONSTERS: *Who's Who Of The Elite, Robert Gaylon Ross Sr.*
When a nation or other: *Political Ponerology, Andrew M. Lobaczewski*
The actions of this phenomenon affect: *Ibid*

The Hidden Evil
If someone has personally experienced: *Ibid*
perverted sense of loyalty: *MHME, Julianne McKinney*
Sadistic control is also an element of perversion: *The Psychopathic Mind, Dr. J. Reid Meloy*
treat their targets as objects: *The Investigative Journal, Julianne McKinney*
Psychopaths view people: *Without Conscience, Dr. Robert D. Hare*
The outer layer closest to the original: *Political Ponerology, Andrew M. Lobaczewski*
According to Dr. Lobaczewski: *Ibid*
envy and greed in themselves: *The Psychopathic Mind, Dr. J. Reid Meloy*

-The Satanic Influence
people who have been stalked after breaking ties: *McKinney MHME*
get rewarded for their evil actions: *Microwave mind control, Dr. Kilde*
License plates which include the numbers: *Personal Notes of Mark M. Rich, www.TheHiddenEvil.com*
Use of modified license plates: *McKinney MHME*
Satanists usually appear: *Techniques Used To Silence Critics, Ted L. Gunderson; The Franklin Cover-up, Senator John Decamp; Satanic Crime, William H. Kennedy; The New Satanists, (Grand Central Publishing, 1994), Linda Blood*
evil needs to hide: *People Of The Lie, Dr. M. Scott Peck*
one of the blossoms and one of the roots: *Ibid*

Satanism encompasses: *Techniques Used To Silence Critics, Ted L. Gunderson; The Franklin Cover-up, Senator John Decamp; Satanic Crime, William H. Kennedy*
of a wider criminal continuum: *The New Satanists, Linda Blood*
The video Conspiracy of Silence: *Techniques Used To Silence Critics, Ted L. Gunderson*
Satanists control and dominate the big corporations: *Ibid*
specific organizations under a Satanic influence: *Ibid*

The Order of Skull and Bones
The Order was founded in 1832: *Rule By Secrecy, Jim Marrs*
It is said to be an American branch: *Ibid*
during the 2004 presidential candidacy: *Baltimore Sun, Keepers of the crypt, March 23, 2004; CBS News, Skull And Bones, June 13, 2004 http://www.cbsnews.com/stories/2003/10/02/60minutes/printable576332.shtml; Washington Post Staff Writer, Bush, Kerry Share Tippy-Top Secret, April 4, 2004, Don Oldenburg; Guardian Unlimited, Skeleton key to the White House, February 24, 2004; The Atlanta Journal-Constitution, Yalies Bush, Kerry can keep a secret (Skull and Bones), March 6, 2004, Bob Dart*
Meet the Press: *www.infowars.com/print/Secret_societies/kerry_bush_sb.htm*
Each year during commencement week: *America's Secret Establishment, Professor Antony C. Sutton*
Members of The Order are elevated: *Ibid*
It was legally incorporated: *Ibid*
It conducts yearly meetings: *Ibid*
These families include, Whitney, Lord: *Ibid*
Bush and Kerry are only the latest Bonesmen: *The Atlanta Journal-Constitution, Yalies Bush, Kerry can keep a secret (Skull and Bones), March 6, 2004, Bob Dart*
Notable members of The Order include: *America's Secret Establishment, Professor Antony C. Sutton*
Columnist William F. Buckley: *Dark Majesty, Professor Texe Marrs*
Bohemian Club member lists: *Bohemian Grove: Cult Of Conspiracy, (iUniverse, 2004), Mike Hanson*
George H. W. Bush was a member of The Order: *America's Secret Establishment, Professor Antony C. Sutton*
George Bush was a Skull and Bonesman: *The Shadows Of Power, James Perloff*
Prescott Sheldon Bush was a founding: *Dark Majesty, Professor Texe Marrs*
Both George H. W. Bush and George: *Bohemian Grove: Cult of Conspiracy, Mike Hanson*
Winston Lord became the chairman: *Americas Secret Establishment, Professor Antony Sutton; Dark Majesty, Professor Texe Marrs*
Henry Lewis Stimson was also a CFR: *The Shadows Of Power, James Perloff*
E. Roland Harriman was a CFR: *Ibid*
W. Averill Harriman, who financed the Nazis: *Ibid (Perloff); Wall Street And The Rise Of Hitler, Professor Antony C. Sutton; Americas Secret Establishment, Professor Antony C. Sutton*
In more recent years we find that during the 2004: *Americas Secret Establishment, Professor Antony Sutton*
The Order has been called a "stepping stone": *Rule By Secrecy, Jim Marrs*
the world shadow government owned: *The Bilderberg Diary, James P. Tucker*

Even with our limited knowledge: *America's Secret Establishment, Professor Antony C. Sutton*
the shortest senior is appointed "little devil,": *The History Channel History's Mysteries, Secret Societies, hosted by Arthur Kent (www.historychannel.com)*
The reader may consider this juvenile: *America's Secret Establishment, Professor Antony C. Sutton*
I can only tell you that this is exactly: *Dark Majesty, Professor Texe Marrs*
It is a psychological conditioning process: *Ibid*
a variation of brain-washing: *America's Secret Establishment, Professor Antony C. Sutton*
Like their counterparts: *Dark Majesty, Professor Texe Marrs*
been able to gain impressive positions of influence: *Ibid*
has infiltrated the top of public organizations: *America's Secret Establishment, Professor Antony C. Sutton*
One observation is that The Order: *Ibid*
infiltrated every aspect of society: *Dark Majesty, Professor Texe Marrs*
The Order being first on the scene is evident: *America's Secret Establishment, Professor Antony C. Sutton*
James Jeremiah Wadsworth: *Ibid*
Andrew Dickson White: *Ibid*
The activities of The Order are directed: *Ibid*

Bohemian Grove
invited to the summer encampment of 1994: *Sonoma County Free Press, San Francisco Bohemian Club: Power, Prestige and Globalism, June 8, 2001, Peter Phillips*
They produced a film of their experience: *Dark Secrets Inside Bohemian Grove, Alex Jones, Mike Hanson, (www.infowars.com)*
Mike Hanson authored a book: *Bohemian Grove: Cult of Conspiracy, Mike Hanson www.cultofconspiracy.com*
a few mainstream media sources: *The Washington Post, Lunch Among The Redwoods, July 06, 1905; New York Post, Gay Porn Star Services Bohemian Grove Members, July 22, 2004; Sacramento Bee Correspondent, Movers, shakers from politics, business go Bohemian, August. 2, 1999, Suzanne Bohan (Infowars.com)*
known as the Midsummer Encampment: *Bohemian Grove, Mike Hanson*
was formed in 1872-3: *San Francisco Bohemian Club, Peter Phillips; Bohemian Grove, Mike Hanson*
invitation only; *Ibid (Hanson)*
According to Hanson and Phillips: *Ibid*
good-old-boys political network *Ibid)*
they rented Bohemian Grove: *Ibid*
While the Grove's public areas: *Ibid*
fire department and medical care: *Ibid*
No Trespassing signs cover the grounds: *Inside Bohemian Grove: Masters of the Universe Go to Camp, November, 1989, Philip Weiss*
barbed wire fences: *Ibid*
vigorously guarded: *Bohemian Grove, Mike Hanson*
own private security force: *Ibid*
Professor Marrs' research: *Exposé of the Bohemian Grove, Professor Texe Marrs interviews Alex Jones www.texemarrs.com*

lasts for about two weeks: *Bohemian Grove, Mike Hanson*
alcohol is available 24/7: *Ibid*
average club member's age: *Inside Bohemian Grove, Philip Weiss*
Between two and three: *Ibid*
The second Saturday of the encampment: *Ibid; Bohemian Grove, Mike Hanson*
Spring Jinks in June: *Ibid (Weiss)*
The CFR and Bohemian Club: *Bohemian Grove, Mike Hanson*
In attendance are representatives: *Extra! Magazine, Inside Bohemian Grove: The Story People Magazine Won't Let You Read (Fairness and Accuracy in Reporting (FAIR)), November/December 1991, www.fair.org/extra/best-of-extra/bohemian-grove.html*
Here, heads of state and industry: *Bohemian Grove, Mike Hanson*
The remaining members are mostly: *San Francisco Bohemian Club, Peter Phillips*
I found at least 150 Bohemians: *Bohemian Grove, Mike Hanson*
Notables that have attended the grove: *Ibid*
consists of a variety of camps: *Ibid*
The Bushes call the Hill Billies: *Ibid*
There are few rules: *Inside Bohemian Grove, Philip Weiss*
The club's motto, was taken: *The Daily Reveille (Louisiana State University), An Elite Alliance, March 7, 2006, Amy Brittain*
The media figures attending: *Extra! Magazine, Inside Bohemian Grove*
Hanson's experience is similar: *Bohemian Grove, Mike Hanson*
Media access to the Grove: *The Daily Reveille, An Elite Alliance*
policy passed without the voting public's knowledge: *Bohemian Grove, Mike Hanson*
begin at 12:30: *Ibid*
The club has drawn criticism: *Sacramento Bee Correspondent, Movers, shakers from politics, business go Bohemian, August. 2, 1999, Suzanne Bohan (Infowars.com)*
The point of the protests: *Ibid*
The important men come out for the Lakeside Talks: *Inside Bohemian Grove, Philip Weiss*
that the Manhattan Project: *Bohemian Grove, Mike Hanson*
Grovers privately boast: *New York Post, Gay Porn Star Services Bohemian Grove Members, July 22, 2004*
the Manhattan Project was conceived on its grounds: *Extra! Magazine, Inside Bohemian Grove*
not an isolated case: *Bohemian Grove, Mike Hanson*
Stealth B-2 Bomber: *Ibid*
Other Lakeside Talk topics: *Ibid*
These speeches were delivered: *Ibid*
Behind the scene: *San Francisco Bohemian Club, Peter Phillips*
These guys ALREADY run the world: *Bohemian Grove, Mike Hanson*
Club members collect owl statues: *Ibid*
decorated with wooden or stone owl statues: *Inside Bohemian Grove, Philip Weiss*
"The Owl" shuttle: *Bohemian Grove, Mike Hanson*
a mock-human sacrifice: *Ibid*
In a letter regarding: *Ibid*
perform mock-druidic rituals: *New York Post, Gay Porn Star Services Bohemian Grove Members, July 22, 2004*
The annual gathering near the Russian River: *Bee Correspondent, Movers, shakers from politics, business go Bohemian*

Grove members carrying torches: *The Daily Reveille, An Elite Alliance*
The cremation took place: *Inside Bohemian Grove, Philip Weiss*
Built to serve as a ceremonial: *Bohemian Grove, Mike Hanson*
Beethoven's 7th symphony: *Ibid*
the sacrificial object is screaming in pain: *Ibid*
silly if it weren't so evil: *Exposé of The Bohemian Grove, Professor Texe Marrs interviews Alex Jones www.texemarrs.com*
owl represents Moloch: *Bohemian Grove, Mike Hanson*
there is no shortage of evidence: *Ibid*
surrounded by men groaning: *Exposé of The Bohemian Grove, Professor Texe Marrs*
Revulsion is a powerful emotion: *People Of The Lie, Dr. M. Scott Peck*
really sick to be inside there: *Exposé of The Bohemian Grove, Professor Texe Marrs*
a Satanic ritual: *Ibid*
There have been rumors circulating: *Bohemian Grove, Mike Hanson*
occult experts analyze their footage: *Ibid*
forced to witness and engage in Satanic activity: *The Franklin Cover-up, former Senator John W. DeCamp*
wealthy individuals were caught: *Blood Sacrifice and Debauchery at the Bohemian Grove-Shocking New Revelations, Professor Texe Marrs interviews Alex Jones www.texemarrs.com*
a snuff pornography film: *The Franklin Cover-up, John W. DeCamp*
the ritual rape and murder: *Bohemian Grove, Mike Hanson*
participate in Satanic cult murder: *Blood Sacrifice and Debauchery at the Bohemian Grove, Professor Texe Marrs*
Male and female prostitutes: *Bohemian Grove, Mike Hanson*
In the fall of 1992: *Ibid*
I didn't know that this place he describes: *Ibid*
sudden deaths: *Blood Sacrifice and Debauchery at the Bohemian Grove, Professor Texe Marrs*
O'Brien writes of her visits to the Bohemian Grove: *Bohemian Grove, Mike Hanson*
In her documented autobiography: *Trance Formation of America, Cathy O'Brien, Mark Phillips*
Even more difficult to ignore: *Bohemian Grove, Mike Hanson*
those who are evil: *People Of The Lie, Dr. M. Scott Peck*
those ushering in the New World Order: *Trance Formation of America, (Reality Marketing Incorporated, 1995), Cathy O'Brien, Mark Phillips*
takes away many of our Constitutional rights: *Techniques used to silence critics, Ted L. Gunderson FBI Senior Special Agent In Charge (Ret)*

Run by Satanists

warned of an evil, one-world government: *The Shadows Of Power, James Perloff*
Eleanor White proclaimed that: *Road to Freedom Show 19, Eleanor White interviews William H. Kennedy, www.shoestringradio.net*
warning them of paranoid extremists: *Project Megiddo, FBI*
I proudly served in the FBI: *Techniques Used To Silence Critics, Ted L. Gunderson*

-Techniques to Discredit
Anyone in the U.S. who: *Trilaterals Over Washington, Professor Antony C. Sutton, Patrick M. Wood*
Propaganda and the Alert Citizen: *The Shadows Of Power, James Perloff*
Because the Establishment controls the media: *None Dare Call It Conspiracy, Gary Allen*
Smear tactics: *Trilaterals Over Washington, Professor Antony C. Sutton, Patrick M. Wood*
phony Left-Right political spectrum: *America's Secret Establishment, Professor Antony C. Sutton*
divert attention from responsible reasoned criticism: *Trilaterals Over Washington, Professor Antony C. Sutton, Patrick M. Wood*
If you assemble the evidence: *The Rockefeller File, Gary Allen*
professional scandalmongers: *With No Apologies, Senator Barry Goldwater*
Barry Goldwater was the Republican: *The Shadows Of Power, James Perloff*

Investigations into Crimes and Practices of the Financial Elite are Thwarted—Witnesses Discredited
appear in the Omaha World Herald: *The Franklin Cover-up, Senator John DeCamp*
Carefully Crafted Hoax: *Ibid*
spawning swirls of gossip: *Ibid*
anonymous complaint: *Ibid*
The cover-up of a similar event occurred in Minnesota: *Ibid*
international organized crime syndicate: *Ibid*
kidnapping ring in Belgium: *Satanic Crime: A Threat In The New Millennium, (www.lulu.com), William H. Kennedy*

The Hidden Evil is Beyond a Congressional Investigation
Wormser described: *Foundations: Their Power and Influence, Rene Wormser*
The far-reaching power of the large foundations: *Ibid*
death threats, harassment of witnesses: *How The World Really Works, Alan B. Jones*
The pressure against Congressional investigation: *Foundations: Their Power And Influence, Rene Wormser*
As we have seen: *The Invisible Government, Dan Smoot*
In 1980, the American Legion: *Rule by Secrecy, Jim Marrs*
died in a curious incident: *Ibid*

The Dangers of a Congressional Investigation
the Reece Committee was also killed: *Foundations: Their Power And Influence, Rene Wormser*
every obstacle which could be put in our way: *Ibid*
led by members of the CFR: *The Shadows Of Power by James Perloff; www.infowars.com/print/Sept11/cfr_whitewash_commission.htm*

The United Nations
the United Nations is a creation of the CFR: *The Shadows Of Power, James Perloff*
All American advocates of supra-national: *The Invisible Government, Dan Smoot*
By 1945, the Rockefellers: *The Rockefeller File, Gary Allen*
Of the American delegates: *The Shadows Of Power, James Perloff*
Nelson and his brothers donated: *The Rockefeller File, Gary Allen*

an outgrowth of the old League of Nations: *Rule By Secrecy, Jim Marrs*
front for the banking elites: *How The World Really Works, Alan B. Jones*
the U.S. Committee for the UN: *The Invisible Government, Dan Smoot*

Project Megiddo
strategic assessment of the potential for domestic terrorism: *Project Megiddo, FBI, http://permanent.access.gpo.gov/lps3578/www.fbi.gov/library/megiddo/megiddo.pdf*
The NWO conspiracy theory: *Ibid*
has been used since the mid 30s: *Chronological history of the new world order, Dennis L. Cuddy, PhD, www.currentstateofaffairs.org/newworldorder.html*
In the last several years:
The Age, Pope calls for new world order (Vatican City), January 2, 2004 (http://www.theage.com.au/articles/2004/01/01/1072908854712.htm); San Jose Mercury News (Knight Ridder Newspapers), Neoconservatives push for a new world order, May 4, 2003, Dick Polman; http://www.prisonplanet.com/061103formationnwo.html, Russia, EU can direct formation of new world order-Putin; http://www.prisonplanet.com/290903africannwo.html (Middle East North Africa Financial Network), African Leader Calls For 'New World Order' at UN Meeting; http://www.prisonplanet.com/010903chinanwo.html, (China People's Daily), China's Top Legislator Calls for Establishment of New World Order; Guardian Unlimited, The prime minister's address to British ambassadors in London, January 7, 2003
An active Federal Government Disinformation Program: *Techniques used to silence critics, Ted L. Gunderson FBI Senior Special Agent In Charge (Ret)*
Distrust of the government: *Future War, Dr John B. Alexander*

False Memory Syndrome Foundation
a group of psychiatrists whose mission: *Mass Control: Engineering Human Consciousness, Jim Keith*
mind-control experiments and the military: *Ibid*
a panel of experts which are used to discredit: *Psychic Dictatorship, Alex Constantine*
media attempted to discredit: *Betrayal Trauma, Professor Jennifer J. Freyd*
arrived at the same conclusion: *Trance Formation of America, Cathy O'Brien, Mark Phillips*

Mainstream News is Used to Discredit Targets
the owner of The Washington Post: *Who's Who of the Elite, (RIE, 3rd printing 2002), Robert Gaylon Ross, Sr.*
The idea of a group of people convinced: *The Washington Post, Mind Games, January 14, 2007, Sharon Weinberger*
wacky claims: *Ibid*
there were some abuses that took place: *Ibid*
Not all people who hear voices are schizophrenic: *Ibid*
a professor of psychiatry at Penn State University: *Ibid*
The very "realness" of the voices: *Ibid*

-Why it Remains
Failure of "Establishment" Support Systems.": *MHME, Julianne McKinney*
Congress and state legislators: *The Investigative Journal, Julianne McKinney*
Discrediting by Psychiatrists: *MHME, Julianne McKinney*

This tactic was heavily used in Russia: *The Search For The Manchurian Candidate, Jonathan Marks*
The communists of the USSR: *Dark Majesty, Professor Texe Marrs*
If you go to a medical doctor you do not: *The Investigative Journal, Julianne McKinney*
aware of existing conditions of targeted patients: *MHME, Julianne McKinney*
Nothing I had researched: *Journey Into Madness, Gordon Thomas*
doctors to help destroy enemies of the state: *The People's State, Mary Fulbrook*
Some have allegedly deliberately sabotaged cases: *MHME, Julianne McKinney*
all the defense lawyers and all the judges were part of it: *Stasiland, Anna Funder*
dismissing targeted individuals that report this: *MHME, Julianne McKinney*
the FBI has been contacted by a great number of people: *Ibid*
Writing to the various [federal]: *The Investigative Journal, Julianne McKinney*
Amnesty International Will Not Intervene: *MHME, Julianne McKinney*
the definition of "enemy" increased as time went on: *Stasiland, Anna Funder*
The number of informal collaborators: *The People's State, Mary Fulbrook*
this happened until about 1/3rd of the population was targeted: *http://edition.cnn.com/SPECIALS/cold.war/experience/spies/spy.files/intelligence/stasi.html*
In our evolution, our survival: *Vital Lies, Simple Truths, Dr. Daniel Goleman*
Truths that rattles one's nerves: *Dark Majesty, Professor Texe Marrs*
not recognizing reality: *The Investigative Journal, Julianne McKinney*
If we are descending into a period of darkness: *Dark Majesty, Professor Texe Marrs*
Some who have tried to assist targeted people: *MHME, Julianne McKinney*
They don't want to lean the truth: *Dark Majesty, Professor Texe Marrs*

-Raising Awareness
it may be too late: *Dark Majesty, Professor Texe Marrs*
The invisible government: *The Invisible Government, Dan Smoot*
The dirty little secret: *http://quotes.liberty-tree.ca/quote/robert_reich_quote_5755*
one sure and final way: *The Invisible Government, Dan Smoot*
[to shut down the Federal Reserve]: *How The World Really Works, Alan B. Jones*
repealing the 1947 National Security Act: *Access Denied: For Reasons of National Security, Cathy O'Brien and Mark Phillips*
withdraw from all international, governmental: *The Invisible Government, Dan Smoot*
The best-known instruments: *How the World Really Works, Alan B. Jones*
The evil hate the light: *People of the Lie, Dr. M. Scott Peck*
The one thing these conspirators: *None Dare Call It Conspiracy, Gary Allen*
their repugnant activities: *Dark Majesty, Professor Texe Marrs*
Unless you are an Insider: *None Dare Call It Conspiracy, Gary Allen*
realize that they have been conned: *Ibid*

-Websites

Gang Stalking and Directed Energy Weapons
www.petermooring.nl
www.raven1.net
www.stopeg.com
www.stopcovertwar.com
ww.shoestringradio.net
www.mindjustice.org
www.thchiddcncvil.com

New World Order
www.wethepeoplefoundation.org
www.takebackwashington.com
www.washingtonyourefired.com
www.freedomtofascism.com
www.davidicke.com
www.911sharethetruth.com
www.911truth.org
www.911busters.com

Independent Media
www.cuttingthroughthematrix.com
www.infowars.com
www.gcnlive.com
www.republicbroadcasting.org
www.breakfornews.com
www.blacklistednews.com
www.wtprn.org

Index